GENETICS and RANDOMNESS

GENETICS and RANDOMNESS

Anatoly Ruvinsky

CRC Press
Taylor & Francis Group
Boca Raton London New York

CRC Press is an imprint of the
Taylor & Francis Group, an **informa** business

CRC Press
Taylor & Francis Group
6000 Broken Sound Parkway NW, Suite 300
Boca Raton, FL 33487-2742

First issued in hardback 2017

ISBN 13: 978-1-138-43010-5 (hbk)
ISBN 13: 978-1-4200-7885-5 (pbk)

Library of Congress Cataloging-in-Publication Data

Ruvinsky, Anatoly.
Genetics and randomness / Anatoly Ruvinsky.
p. ; cm.
Includes bibliographical references and index.
ISBN-13: 978-1-4200-7885-5 (pbk. : alk. paper)
ISBN-10: 1-4200-7885-2 (pbk. : alk. apper)
1. Genetics. 2. Stochastic processes. 3. Natural selection. 4. Variation (Biology) I. Title.
[DNLM: 1. Genetic Variation. 2. Evolution, Molecular. 3. Genetic Drift. 4. Recombination, Genetic. 5. Uncertainty. QU 500 R983g 2010]

QH430.R88 2010
576.5--dc22 2009014187

Visit the Taylor & Francis Web site at
http://www.taylorandfrancis.com

and the CRC Press Web site at
http://www.crcpress.com

Contents

...Science, if it is to flourish, must have no practical end in view.

Albert Einstein

The World As I See It

Preface

Molecular genetics, born in the middle of the twentieth century, advanced immensely during the following decades. The basic questions about organization of life on the molecular level were nearly resolved. Genomes of many species have been sequenced and opportunities for new discoveries look great. This spectacular progress in unraveling the most fundamental principles of living matter led to an array of practical applications in medicine, numerous industries, and agriculture. Intellectual power produced by this impressive development seems unlimited and further advances follow with unprecedented speed. Rightly or wrongly, human reaction to these developments varies from admiration to rejection. Expressions like "genetics plays god" have become common and some countries have introduced laws prohibiting use of a wide range of molecular genetic technologies.

Neither science nor society had enough time to fully comprehend the scale of new biological revolution, and deep considerations of this new paradigm are still emerging. In some ways there are similarities between the current situation in genetics and the circumstances in theoretical physics about seventy years ago. In both cases science has reached certain limits and faced an unfamiliar world, sometimes beyond its grasp.

A significant degree of randomness typical of life has been known for a long time. Brilliant scientist and Nobel Prize laureate J. Monod wrote in his book *Chance and Necessity* more than thirty-five years ago: "Randomness caught on the wing, preserved, reproduced by the machinery of invariance and thus converted into order, rule, necessity." Similar understanding of the essence of life and evolution can be found in other publications. The origin of such views can ultimately be found in the Darwinian system where random hereditary changes are considered as fundamental elements of biological evolution. In this book we attempt consideration of randomness in major genetic processes and events.

Chapter 1 is an introduction to the problem and it summarizes the major ideas relevant to uncertainty and randomness in physics and mathematics. It is crucial to realize that randomness is universal and a very basic phenomenon that has few different physical sources as well as some roots in logic. Not surprisingly biology provides countless evidence of randomness, which can be observed on all levels of biological organization, including genes and cells, organisms, populations and the evolutionary process as a whole. In this book we are going to explore the nature of genetic uncertainty from different points of view and at different levels. Chapter 2 is devoted to mutations. The proposal that many mutations are spontaneous and probably related to or even caused by subatomic randomness became popular among a group of physicists, including E. Shrödinger and M. Delbrück, well before the discovery of DNA structure. Here we shall discuss this fundamental problem again in order to trace the genesis of these ideas and also the deep links between subatomic fluctuations and long-term macroscopic changes in living organisms. Such understanding of the nature of mutations was strongly confirmed by numerous investigations and now can be found in textbooks.

Chapter 3 considers recombination events as the second layer of randomness that contributes greatly to genetic variability. The number of combinations resulting from constantly ongoing recombination processes is huge. This provides a practically inexhaustible source of new genetic variations. A multitude of random events that occur during development is the subject of Chapter 4. One of the reasons for the vast gap between phenotype and genotype is developmental randomness. Filtering and utilization of this type of randomness is essential in the process of adaptations building. Even more than that, randomness became a core phenomenon in immune response at least in high vertebrates. An ability to generate randomness on a grand scale becomes a matter of life and death. One can speak in this case about "promoted" randomness.

Chapter 5 concentrates on segregation, another level of genetic processes found in diploid organisms. Here randomness is completely organized and reaches a maximum degree of uncertainty. A chance for nearly each allele to be transmitted to a particular gamete is about 50%. Only in very rare situations can specific genes and chromosome structures cheat the mechanism of equal segregation and thus lead to segregation distortions. As Chapter 6 shows, the majority of if not all population processes include a significant degree of randomness. This was clearly understood by the founders of population genetics and became an even more prominent part of the theoretical framework in the last decades. The cornerstones of evolutionary process, genetic drift and fluctuating natural selection, are the best demonstrations of randomness at this level of biological organization.

The major aim of science is finding general principles, laws, and mechanisms, as well as regular and predictable events. Randomness, at first glance, is the alternative to all of these and may as appear to be an annoying nuisance that limits science and prevents generalizations. Such a deterministic and outdated understanding of science is steadily being replaced by the probabilistic view which became so typical for genetics. All these issues are discussed in Chapter 7. Hopefully this book will be helpful in the consolidation and propagation of this probabilistic understanding of life.

It is my pleasant duty to acknowledge the great support I have received from the University of New England, Australia. This book would hardly have been written without the generous sabbatical and other help provided by the university. I am also in debt to many colleagues and friends with whom I discussed different aspects of randomness over the years. I am grateful to Brian Kinghorn for discussions and his comments on the manuscript. Certainly, warm thanks go to my family for the constant help.

Anatoly Ruvinsky
Coffs Harbour and Armidale
New South Wales

chapter one

Limits and uncertainty in nature and logic

Every science, as its observations accumulate and its paradigms complexify, may be expected to approach the qualitative limit of relationality. … But it is also a limit sciences must refuse to cross if they are to remain scientific.

Brian Massumi
Parables for the Virtual
(2002)

Limits of nature

Scientific curiosity seems unlimited. During the last couple of centuries a myriad of questions and problems have been raised and solved. This has created the impression that science can eventually find an answer to any question and transform the unknown of yesterday into the well understood of tomorrow. Perhaps this is true for many scientific problems but hardly for all. The most fundamental scientific discoveries of the past also unraveled some limits of matter beyond which nothing makes sense. Such limits as the speed of light, absolute zero, and quantum of energy are familiar to all who read textbooks. The physical limits were initially predicted using a theoretical approach and only later were tested in experiments. These limits stand like absolute taboos in the way of gaining further knowledge.

About 2500 years ago ancient Greek thinkers deduced the first fundamental limit of nature. They put forward the idea of the discrete structure of matter. Only in the nineteenth and early twentieth centuries was this idea finally supported by hard experimental data, and the term *atom* became commonly used. Although the theory of the indivisibility of atoms did not hold for too long, the atomistic idea is alive. Erwin Schrödinger (Figure 1.1 and brief biographic note), in his excellent essay "Science and Humanism" (1951), provided the most interesting interpretation of this deep philosophical conclusion. He rejected the notion that the ancient thinkers just luckily came to such a deep realization. Instead Schrödinger believed that this idea was a result of intellectual defense

Figure 1.1 Erwin Schrödinger (1887–1961), an outstanding Austrian physicist and Nobel laureate (1933). He made a great contribution to quantum mechanics and also to theoretical biology by writing the famous book *What Is Life?* He was born and died in Vienna. He worked in Zürich and Berlin and then for many years at the Institute of Advanced Studies, Dublin University, Ireland. (Courtesy of the *Dublin Institute for Advanced Studies and the Irish Time.*)

against the overly powerful tyranny of the mathematical continuum. The mystery of continuum is that an infinite amount of points or numbers can be inserted between say 0 and 1 or any other two arbitrary points or numbers. Essentially the interval between 0 and 1 can be stretched infinitely. Regardless of how this question is resolved in mathematics, in the physical world the number of material components in any given volume usually does not increase, if a piece of matter is stretched further and further. Such logic required the invention of indivisible elements called "atoms" which filled a particular space. Although during the last several decades the simple atomistic concept was eventually replaced by much more sophisticated theory, the core idea of the early atomists still holds and presumes the existence of indivisible elements, which, at different times, were associated with subatomic particles, quanta, strings, etc. Such indivisible elements, in spite of what we call them, represent the smallest components of matter or minimal quantity of energy and thus indicate the fundamental limit of nature. The question of what is below this level does not make sense and further scientific investigation is impossible. This became particularly relevant for physics in the twentieth and twenty-first centuries. It seems likely that the spectacular progress in physics for the last hundred years has brought it to the very limits of the knowable. A critical account of the current situation was described recently by Smolin (2007). Obviously science will never stop but it would probably be too optimistic to think that there is no limit to its progress.

Another fundamental limit is absolute zero. As early as in 1702 French physicist Guillaume Amontons came to the conclusion that -270° on the Celsius scale is the lowest possible temperature. He thought that the spring of air in his air thermometer must be reduced to nothing at this temperature. Subsequent theoretical progress discovered that -273.15°C is the absolute temperature limit; a system neither emits nor absorbs energy when it reaches this temperature. In reality this temperature cannot be reached but the latest result obtained by physicists from the Massachusetts Institute of Technology was exceptionally close to absolute zero (http://en.wikipedia.org/wiki/Absolute_zero). Matter, as we know it, cannot be cooler than absolute zero. Again as in the case of atoms, it is impossible to understand what is below this magic temperature and the question is senseless, at least within the framework of modern science.

Now it is common knowledge that the speed of light in a vacuum is about 300,000 km/sec and a material object or information cannot travel faster. If information could travel faster than the speed of light in one reference frame, causality would be violated and in some other reference frames the information would be received before it had been sent. This means that the "result" could be observed before the "cause" (http://en.wikipedia.org/wiki/Speed_of_light). One hundred years ago, when Albert Einstein pronounced his views, it was probably hard news to swallow. Despite the powerful theoretical taboo there were many attempts to overcome this limit, at least terminologically. Here we shall not discuss these experiments although some were very sophisticated and even spectacularly successful. The reason behind this is a vague definition, which essentially defined light as a combination of several processes. This fundamental limit of nature beyond the fact that it has deep theoretical significance might become very important from a practical point of view. For instance, the speed of light may eventually develop into an obstacle for further increase of the speed of computers because it is a critical limit for the information exchange between computers. More recently a theoretical possibility was considered that the speed of light might change during the evolution of the universe (Smolin 2007). Nevertheless, as far as I know, there is no debate about the limit itself.

In this short introduction it is difficult to avoid a powerful example of the time-related limit. Much data indicates that the universe is a result of the Big Bang, which took place approximately 13.7 billion years ago. Prior to this incredible event the universe and hence time did not exist. This phenomenon is quite perplexing and represents another mystery which will never be finally resolved by scientific means. The existence of the aforementioned absolute limits of the material world cannot be explained by lack of knowledge; on the contrary they represent the finest achievements of science. These limits rather demonstrate "strange" laws of nature, which are not easy to understand and impossible to overcome. In general we probably should not anticipate that science can provide more than an adequate reflection of the material world and generate reasonable expectations of earlier unobserved or forthcoming phenomena. Certainly it is not the same as finding an absolute Truth. This less than optimistic notion of science appeared in the early twentieth century when quantum mechanics was developing. As I have already mentioned, the remarkable progress in physics led to discovery of these fundamental limits and as a result physics was the first branch of science to face "strange" realities on the edges of matter. Other fields of science, as soon as they reach a certain degree of development, should face similar challenges and genetics is one of them.

Quantum uncertainty principle

Quantum theory symbolizes a dramatic split from the classical tradition in physics. Werner Heisenberg (Figure 1.2 and brief biographic note), who was invited by Niels Bohr to be a research associate in his Copenhagen Institute, suggested an entirely different approach for describing the dynamics of subatomic particles. In 1925 he wrote: "My entire meagre efforts go toward killing off and suitably replacing the concept of the orbital paths that one cannot observe." Then, he was only 24 years old. Heisenberg replaced the complexities of three-dimensional orbits of subatomic particles by a one-dimensional vibrating system. The result of his efforts was the famous formulae in which quantum numbers were related to observable radiation frequencies and intensities. A year later he formulated the uncertainty principle, thereby laying the foundation for a new interpretation of quantum mechanics, and a few years later he was awarded the Nobel Prize. The essence of this central statement of quantum mechanics in a simple form is that it is impossible to simultaneously know or measure the position and speed of an elementary particle. This is always characterized by a probability distribution. In his 1927 uncertainty paper Heisenberg wrote: "The more precisely the position is determined, the less precisely the momentum is known in this instant, and vice versa" (http://www.aip.org/history/heisenberg/p08.htm).

Figure 1.2 Werner Heisenberg (1901–1976), an outstanding German physicist and Nobel laureate (1932), one of the founders of quantum mechanics. He was born in Würzburg and died in Munich Bavaria, Germany. (Photo courtesy of Werner-Heisenberg-Institut, Munich.)

Soon after, in 1930, Percy Robertson and Erwin Schrödinger independently developed a generalization of the uncertainty principle, which led to the Nobel Prize being awarded to Schrödinger. In the following two decades a few more important extensions of the principle were discovered between energy and position, energy and time, etc. (http://en.wikipedia.org/wiki/Uncertainty_principle).

Thus, the uncertainty principle denounced a possibility of exact simultaneous knowledge of basic characteristics of subatomic particles, which led to the realization that complete and precise description of matter is impossible. The deterministic approach, which dominated science for a long time, was shaken and even became inadequate in this strange quantum world. Randomness took a much more prominent position in physics and probabilistic description became the mathematical tool of choice. Then Niels Bohr and Werner Heisenberg made another step: they ruled out even a theoretical possibility to obtain "in principle" complete knowledge about a macroscopic object, for example an individual. This view sprang from their belief in intricate interactions between subject and object. According to this vision a very strong and deep intervention is required in order to get so-called maximal description of an object, which essentially means that the object has to be eventually destroyed in the process of observation and collection of essential information. Schrödinger did not entirely share this view and stated that nothing in contemporary quantum theory in principle prevents an opportunity to obtain complete knowledge about a living creature. However, he stressed the fundamental difference between classical and quantum predictions following from the maximal description of an object: uncertainty in behavior of the object, described in quantum terms, increases significantly as time goes by and hence the prediction becomes fuzzy. More recently W. H. Zurek (1998) has shown "that the evolution of a chaotic macroscopic (but, ultimately, quantum) system is not just difficult to predict (requiring an accuracy exponentially increasing with time) but quickly ceases to be deterministic in principle as a result of the Heisenberg indeterminacy (which limits the resolution available in the initial conditions)."

The departure from the deterministic view of nature was not easily accepted and quite expectedly caused significant controversy and disagreement. Probably the most vivid demonstration of such differences in opinion was the unwillingness of Albert Einstein to accept the uncertainty principle and hence the new interpretation of quantum mechanics. As the great thinker of our time famously put it, "I cannot believe that God would choose to play dice with the universe." It must be stressed that while the exact characteristics of a single elementary particle are not known, a highly precise description of a beam consisting of millions of particles is possible. This gives an optimistic outlook on the explanatory power of science but it may lead to ignoring the relevance of the uncertainty principle for macroscopic objects. Due to a gigantic difference between the size of a subatomic particle and a macroscopic object, it is conventional to consider influences of the former on the latter as infinitely small. Generally this might be correct but in a number of biological processes this is not exactly so. One of the objectives of this book is to show that uncertainty generated on the subatomic level is crucial for explaining some basic genetic processes. As Chapter 2 demonstrates numerous individual genetic events are unpredictable in principle; however, a stream of such events can be predicted reasonably well using probabilistic methods.

Further penetration of these important physical concepts in biological thinking and in contemporary theories might be fruitful and needs special consideration. As we shall see in the following chapters, randomness generated on the subatomic level is very much relevant to many phenomena and processes observed in living organisms and should not be ignored.

Statistical mechanics and Brownian motion

In complex physical systems there is another source of uncertainty. This was realized long ago and many great minds like Ludwig Boltzmann, James Maxwell, Max Planck, and Albert Einstein, to name just a few, contributed to this field of theoretical physics. Let us assume for the sake of argument that movements of atoms or molecules in a gas are completely deterministic and individual trajectories can be calculated in principle. Nonetheless it seems absolutely impossible to actually describe or predict all individual trajectories in a bottle of gas as this would require knowledge of an incalculable number of initial conditions ($>10^{20}$). It is only one of the insurmountable problems in the way of obtaining comprehensive knowledge of a macroscopic system. A solution to the problem was steadily developed in thermodynamics by introducing general macroscopic parameters such as entropy, energy, temperature, etc., and creating the necessary mathematical tools. Using these parameters an accurate description of the macroscopic system is possible despite a complete lack of information about the movements of individual elements that comprise the system. This was an important methodological shift, which allowed dealing with macroscopic objects whose inner structure and dynamics are unknown. Further progress in this field is related to statistical mechanics, which is capable of making macroscopic predictions based on microscopic properties. The predictions or descriptions made in the frame of statistical mechanics are certainly probabilistic and in that sense it is similar to quantum mechanics. Randomness is an inalienable feature of complex physical systems. Certainly the same is true for all kinds of biological systems, which are very complex and nonlinear; randomness is also their core characteristic.

Randomness can be easily observed. In 1827 English botanist Robert Brown described erratic movements of pollen grains suspended in water, which were seen under a microscope. He clearly understood that these random movements had nothing to do with pollen itself because even those pollen grains that had been stored in dry conditions for a very long time behaved in exactly the same way. The decisive point in the investigation of the nature of Brownian motion was provided by Albert Einstein. He had shown that random movements of light visible bodies suspended in liquids are caused by the thermal motions of molecules and atoms. A small particle is regularly and randomly bombarded by molecules of the liquid from all sides. The hits coming from one side in a particular moment by chance might be stronger than from another side, and this is the cause of movements. These tiny random jumps create an erratic trajectory typical for Brownian motion. There are many computer programs that simulate the process rather well (Figure 1.3). The importance of Brownian motion is at least twofold. First, this is a simple visual proof of molecular movements; second, it shows that movements of molecules and atoms can affect the dynamics of macroscopic objects. No doubt for large objects like animals and plants such influences are nearly equal to zero. However, for small objects like separate living cells or some cellular components, which can move, Brownian motion is a constant factor making a difference.

Good examples of the successful application of the general methodology developed by statistical mechanics can be found in two classical branches of genetics. The first one is quantitative genetics. The

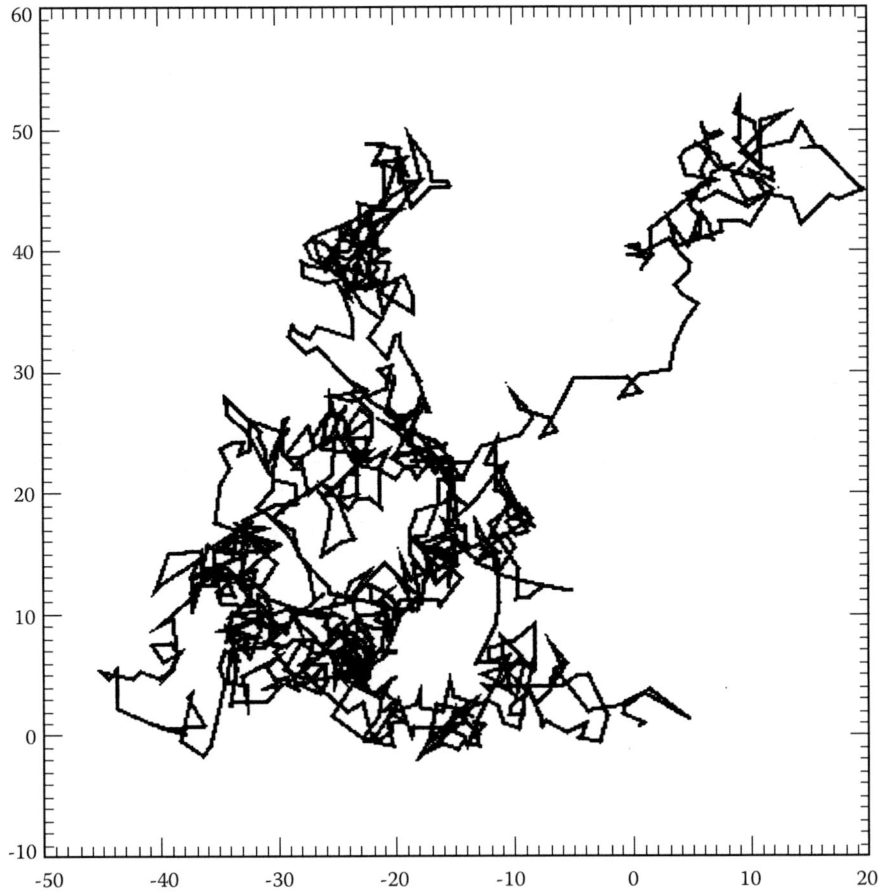

Figure 1.3 An example of Brownian motion in the plane. (Photo from http//upload.wikimedia.org/wikipedia/commons/5/59/BrownianMotion.png.)

founders of this branch of genetics, Ronald A. Fisher, Sewall G. Wright, and John B. S. Haldane, realized the extreme difficulties in connecting phenotypes of individuals with their genotypes for complex traits with continuous variation, so-called quantitative traits, like the height or weight of individuals. There are two major problems here: (1) a lack of sufficient knowledge about the hundreds of genes (alleles) involved in determination of such traits in individuals and the possible complicated interactions between such genes; and (2) the dramatic influence of environmental factors on the activity of all these genes. The classic Mendelian approach is unable to take into account the contributions of numerous separate genes and environmental influences. As an alternative, quantitative genetics developed a statistical methodology in order to tackle this hard issue. A set of parameters, such as heritability, genetic correlations, and breeding values, were introduced, which allow describing complex processes without knowledge of genotypes, countless gene interactions, and environmental influences. Such "deviation" from the standard genetic methodology remains very successful. Another bright example is population genetics. It is remarkable that nearly the same group of scientists developed the foundation of theoretical population genetics. As in the case of ideal gas in physics, a complete knowledge of a large population is practically impossible. However, general parameters such as frequency of alleles, heterozygosity, effective size of population, coefficient of migration, and so on are quite sufficient to give a reasonable description and prediction of the population structure and dynamics. Further discussion of these topics can be found in Chapters 4 and 5. As a result many predictions in quantitative and population genetics are inevitably probabilistic. Although theoretical population genetics has the strongest mathematical foundations, it can describe or predict the structure and dynamics of a finite population only in probabilistic terms. Regardless of the quantity of available information in the future, population genetics will always remain probabilistic, which is the only reasonable way to describe reality.

Obviously the problems that we have briefly outlined in relation to some genetic phenomena are not unique and are highly relevant to vast fields of science, technology, social sciences, and so on. Below is a citation from the editorial page of the interdisciplinary journal of nonlinear science, *Chaos*:

> In the past two decades the "new science," known popularly as "chaos," has given us deep insights into previously intractable, inherently nonlinear, natural phenomena. Building on important but isolated historical precedents (such as the work of Poincaré), "chaos" has in some cases caused a fundamental reassessment of the way in which we view the physical world. For instance, certain seemingly

> simple natural nonlinear processes, for which the laws of motion are known and completely deterministic, can exhibit enormously complex behavior, often appearing as if they were evolving under random forces rather than deterministic laws. One consequence is the remarkable result that these processes, although completely deterministic, are essentially unpredictable for long times. But practitioners of "nonlinear science," as "chaos" has become known among experts, recognize that nonlinear phenomena can also exhibit equally surprising orderliness. (http://chaos.aip.org/chaos/staff.jsp)

Thus, at least two powerful sources of randomness exist in nature: uncertainty of the quantum world and extreme complexity of micro- and macroscopic dynamic systems. Both are highly relevant to genetic processes.

Randomness in mathematics

Most people believe, and with good reason, that mathematics is always precise. Mathematics, which is indeed one of the most ancient human intellectual endeavors, strives to perfection and just until several decades ago most, if not all, mathematicians were absolutely sure that mathematics is boundless in its pursuit of knowledge and elegance and certainly is absolutely precise. Those who are interested in getting a deeper understanding of this serious matter could read the book *Meta Math!* written by Gregory Chaitin (2005). Here I intend to give only a brief sketch of the problem, which is relevant to the major theme of this book.

More than a century ago David Hilbert, the towering figure in mathematics, made a spectacular attempt to create a formal axiomatic system (FAS) for mathematics and by doing this to remove uncertainty in mathematic argumentation and definition. Hilbert is viewed by some mathematicians as the creator of metamathematics, which is a kind of foundation of mathematics. He compelled a set of twenty-three great mathematical problems and believed that each of them will be resolved. Hilbert's views are well expressed in the following statement made in 1900:

> The conviction of the solvability of every mathematical problem is a powerful incentive to the worker. We hear within us the perpetual call: There is the problem. Seek its solution. You can find it by pure reason, for in mathematics there is no *ignorabimus*. (http://www.math.umn.edu/~wittman/hilbert.html)

Hilbert's influence on mathematics in the twentieth century was profound and represented the great achievement of classical mathematics. It came as a surprise in 1931 when Kurt Gödel, then 25 years old (Figure 1.4 and brief biographic note), published his famous paper, which states that for any self-consistent axiomatic system powerful enough to describe the arithmetic of the natural numbers, there are true propositions about the naturals that cannot be proved from the axioms. Two major conclusions of this work are particularly important:

1. If the system is consistent, it cannot be complete. (This is known as the incompleteness theorem.)
2. The consistency of the axioms cannot be proved within the system.

Figure 1.4 Kurt Gödel (1906–1978), one of the most prominent logicians of all time. He worked for many years at Princeton Institute of Advanced Studies. Gödel and Albert Einstein were close friends. Gödel was born in Brünn Austro-Hungary (now Brno, Czech Republic), and died in Princeton. (Courtesy of the Archives of the Institute for Advanced Study, Princeton, NJ.)

The incompleteness theorem implies that not all mathematical questions are computable (http://en.wikipedia.org/wiki/Kurt_G%C3%B6del). In essence this theorem puts impassable limitations on creating logically perfect foundations of mathematics. This work had a great deal of importance for several basic areas of mathematics and philosophy. Limitations discovered by Gödel are purely logical. Reflecting on this logical problem and its consequences Gregory Chaitin (2005) wrote: "The view that math provides absolute certainty and is static and perfect while physics is tentative and constantly evolving is a false dichotomy. Math is actually not that much different from physics. Both are attempts of human mind to organize, to make sense, of human experience."

In 1936, soon after Gödel's discovery, Alan Turing (Figure 1.5 and brief biographic note) made a very important contribution to the foundations of mathematics and future computer science. Prior to the computer era he developed a mathematical concept of the Turing machine,

Figure 1.5 Alan Turing (1912–1957), an outstanding English mathematician and founder of computer science. He was born in London and died in Wilmslow, Cheshire, UK. (Courtesy of the Library Archives of King's College, Cambridge, UK.)

a computational device with unlimited power. The result of his work, which is also relevant to this book, is the famous halting problem. He tried to understand whether an entirely self-contained computer program, which can be provided by any numbers it may need by the program itself, will ever stop. The result obtained by Turing is quite unusual for mathematics. There is no way to know in advance whether the program will halt. It may look in the first instance like a "technical" question; however, this is a very deep mathematical and even philosophical problem. By proving this theorem Turing has shown that any formal axiomatic system (FAS), which can be developed, will not allow answering the "simple" question whether or not the program will ever halt. It means that any FAS is incomplete and this in turn defeats the idea of building an entirely complete and self-sufficient foundation of mathematics. This probably was the first instance in the history of mathematics when classical logical statements like "yes" or "no," 1 or 0 are unachievable in principle and uncertainty is the answer. This line of reasoning indicates that the uncertainty observed in real physical or biological systems might be caused not only by a lack of information but could represent the fundamental feature of such systems. Alan Turing was only 24 years of age at the time.

Gregory Chaitin (2005) made a further step along the path laid out by Gödel and Turing. He was able to show that there is randomness in number theory, the basic field of pure mathematics. Chaitin also indicated that a good demonstration of uncertainty in mathematics is the existence of several mathematical problems that have remained unresolved for centuries, despite numerous attempts to solve them.

It is relatively easy to describe and, in many instances, to predict the following elements of a regular string like 02020202 … 020202 …. The amount of information such a string can carry is quite small regardless of the length due to its high regularity. Things become much more difficult when a random string has to be described. How to measure randomness in this case? Three independent attacks on the problem made by outstanding mathematicians Ray J. Solomonoff in 1964, Andrey N. Kolmogorov in 1965, and Gregory Chaitin in 1969 laid the foundation for the so-called algorithmic theory of randomness. The idea was to describe randomness through complexity, often referred to as Kolmogorov complexity. According to Kolmogorov, the complexity of an object is the length of the shortest computer program that can reproduce the object. Random objects, in this theory, are their own shortest descriptions. As there is no regularity in random numbers, their own descriptions are irreducible and one can talk about the irreducible complexity of random events. This notion of complexity, as a measure of randomness, is closely related to Claude Shannon's entropy of an information source. Starting from the much unexpected conclusions made by Kurt Gödel in 1931, it became increasingly clear that randomness has a place even in pure mathematics. Gregory Chaitin (2006) goes even further:

> Unlike Gödel's approach, mine is based on measuring information and showing that some mathematical facts cannot be compressed into a theory because they are too complicated. This new approach suggests that what Gödel discovered was just the tip of the iceberg: an infinite number of true mathematical theorems exist that cannot be proved from any finite system of axioms.

It seems that uncertainty or in other words randomness is not just a nuisance but a core phenomenon in mathematics and, if so, is a reflection of the way in which logic can operate.

Limits of reasoning: randomness and complexity as the general feature of nature and mind

In this short introductory chapter we have spotted different types of randomness, which sprang either from quantum uncertainty or statistical complexity or human logic. While the sources of these types of uncertainty look unrelated, I am not sure whether this is true. It is not easy to rule out that the different types of randomness are caused in the end by the specific features of human logic. Perhaps there are limits of reasoning beyond which many scientific or mathematical questions cannot be answered in principle. The basic belief that any scientific problem can eventually be resolved, given all necessary information, tools, and logic, might be too optimistic after all.

Chaitin (2006), the rigorous proponent of these emerging yet not conventional views, strongly believes that certain mathematical facts are true for no reason, which is a dramatic deviation from the common logical principles. According to him "an infinite number of mathematical facts are irreducible, which means no theory explains why they are true. These facts are not just computationally irreducible, they are logically irreducible. The only way to 'prove' such facts is to assume them directly as new axioms, without using reasoning at all." As the same logic is relevant to science, one can expect a multitude of natural phenomena which may not be fully explained or predicted. As far as I can see such epistemology[1] does not indicate a potential crisis in science and does not look gloomy.

The enormous progress in different fields of science, including genetics, reached the boundaries within which the classical methodology is working successfully and logical reasoning is the basic principle.

Outside these boundaries a modification of the basic principles seems to be unavoidable and the traditional scientific question "why" may not be answered because a sufficient reason does not exist.

Outstanding physicist and Nobel laureate Murray Gell-Mann (1996), assuming "that the fundamental theory of matter and the initial conditions of the universe are simple and knowable," compiled a list of sources of unpredictability. In addition to those that have been discussed here so far, he also mentioned "unavoidable ignorance," which related to principal impossibility to collect numerous types of information now or at any time in the future. The list also includes the probabilistic character of all branching events or choices which may occur in the future. There are also impassable calculation difficulties especially when histories of events and initial conditions are not certain. He also pointed at so-called amplification mechanisms, which are essential for matter in general. This term refers to a situation in which an outcome is sensitive to the smallest fluctuations in a nonlinear system. Chaos typical for such systems may serve as a source of amplifications. Thus, a tiny deterministic or probabilistic change in the initial or current conditions may lead to a very significant shift in the outcome. Chapter 2 pays particular attention to this mechanism in life processes. The incredible complexity of living organisms has a twofold effect; on the one hand it magnifies the potential for chaos, and on the other hand it promotes mechanisms of self-regulation and self-organization. Evolution of increasingly complex adaptive biological systems is an excellent example of harnessing chaos. Random events that are integral elements of any chaotic system make a decisive impact on the whole process of evolution. Any individual creature not only complies with the fundamental laws of physics but it also represents a result of the unimaginably long chain of probabilistic events. Each of these events could have a different outcome. Obviously any biological organism has an uninterrupted line of ancestors which goes back to the very beginning of life on Earth.

Complexity, randomness, and unpredictability are the basic characteristics of evolving matter. Determinism and predictability are rather ideal, local, and temporary phenomena usually ignoring subtle effects and because of that also unable to make precise predictions of the remote future events. In living matter, the most complex known phenomenon, room for determinism and predictability is very limited. All this creates inescapable impediment for reasoning. Fortunately enough strict limits for further understanding are located at the fringes and because of that science is free to strive toward potential new knowledge with great success and speed. Nevertheless the important emerging conclusion is that both science and logic have absolute limitations. For me this message is an indication of maturity for a simple and profound reason: nothing is endless.

Summary

This chapter serves as a necessary introduction to the problem of randomness and is based on considerations of relevant examples mainly from physics and mathematics. The whole point of this endeavor is the creation of an essential background for numerous manifestations of randomness in genetics. It also serves as a reminder of how strong the similarity is between biological and physical systems. Two powerful sources of randomness are known in nonliving matter: quantum uncertainty and stochasticity of micro- and macroscopic dynamic systems. Both of them are highly relevant to randomness observed in genetics. As the size gap between a subatomic particle and a macroscopic object is huge, it is quite common to ignore quantum effects in microscopic objects. While generally this might be correct, in several biological processes quantum effects play a very important role. In this book we are going to show that uncertainty generated on the subatomic level is crucial for explaining some basic genetic processes. Stochasticity of atoms and molecules is another ubiquitous phenomenon of inorganic matter, which is very relevant to all biological systems and discussed in this chapter. Some great achievements of science established the absolute limits of the knowable. This is true not only for the experimental and theoretical sciences but for mathematics and logic. Proper realization of these important intellectual advancements is very useful for the rapidly developing areas of science and genetics in particular.

Note

1. "Epistemology without contact with science becomes an empty scheme. Science without epistemology is—insofar as it is thinkable at all—primitive and muddled" (Albert Einstein, *The New Quotable Einstein,* collected and edited by Alice Calprice. Princeton, NJ: Princeton University Press [2005], 263).

References

"Absolute zero." Wikipedia. http://en.wikipedia.org/wiki/Absolute_zero (accessed September 4, 2008).

Chaitin, Gregory. 2005. *Meta Math! The quest for Omega*. New York: Knopf.

———. 2006. The limits of reason. *Scientific American* 294 (3):74–81.

"Chaos." http://chaos.aip.org/chaos/staff.jsp (accessed September 4, 2008).

"David Hilbert." http://www.math.umn.edu/~wittman/hilbert.html (accessed September 4, 2008).

Einstein, Albert. 2005. *The new quotable Einstein,* collected and edited by Alice Calprice. Princeton, NJ: Princeton University Press.

Gell-Mann, M. 1996. *Fundamental sources of unpredictability.* Address at conference "Fundamental Sources of Unpredictability," Santa Fe Institute, Santa Fe, AZ. http://www.santafe.edu/research/publications/workingpapers/97-09-079.pdf (accessed December 21, 2007).

"Incompleteness theorem." Wikipedia. http://en.wikipedia.org/wiki/Kurt_G%C3%B6del (accessed September 4, 2008).

Massumi, Brian. 2002. *Parables for the virtual: Movement, affect, sensation.* Durham, NC: Duke University Press.

"Quantum mechanics. The uncertainty principle." http://www.aip.org/history/heisenberg/p08.htm (accessed September 4, 2008).

Schrödinger, Erwin. 1951. *Science and humanism. Physics in our time.* Cambridge, UK: Cambridge University Press.

Smolin, Lee. 2007. *The trouble with physics.* Boston: Houghton Mifflin.

"Speed of light." Wikipedia. http://en.wikipedia.org/wiki/Speed_of_light (accessed September 4, 2008).

"Uncertainty principle." Wikipedia. http://en.wikipedia.org/wiki/Uncertainty_principle (accessed September 4, 2008).

Zurek, W. H. 1998. Decoherence, chaos, quantum-classical correspondence, and the algorithmic arrow of time. *Physica Scripta* T76:186–198.

chapter two

Quantum fluctuations, mutations, and "fixation" of uncertainty

> Almost all aspects of life are engineered at the molecular level, and without understanding molecules we can only have a very sketchy understanding of life itself.
>
> **Francis Crick**
> *What Mad Pursuit*
> *(1988)*

Nature of genes and mutations: the early attempts

Niels Bohr, the great physicist and Nobel laureate, became interested in theoretical problems of biology in the early 1930s. His lecture entitled "Light and Life" and delivered to the International Congress on Light Therapy in Copenhagen in 1932 was not only unusual but had a lasting effect. Young physicist, Max Delbrück (Figure 2.1 and Box 2.1), who was on a postdoctoral Rockefeller fellowship at Bohr's Institute at that time, attended the lecture and was strongly impressed. Bohr's thoughts that the complementarity observed in quantum mechanics might have implications in biology were particularly attractive to Delbrück. In the following months and years Bohr also conducted a series of seminars devoted to basic biological problems. In addition to a dozen bright physicists, including Delbrück, he also invited Nikolay Timofeeff-Ressovsky, a Russian geneticist who had been working in Berlin since 1925 and was interested in theoretical biology (Figure 2.2 and Box 2.2). Soon after, Delbrück returned to Berlin and took a position at the Kaiser Wilhelm Institute for Chemistry. Collaboration between Timofeeff-Ressovsky and Delbrück began.

A third member of the newly formed group was Karl Zimmer, physicist and radiation biologist, and also a member of Timofeeff-Ressovsky's department. He described this collaboration as follows: "Two or three times a week we met, mostly in Timofeeff-Ressovsky's home in Berlin, where we talked, usually for ten hours or more without any break, taking some food during the session. There is no way of judging who learned most by this exchange of ideas, knowledge and experience, but it is a fact

Figure 2.1 Max Delbrück (1906–1981), outstanding German and American scientist, Nobel laureate (1969) for discoveries concerning "the replication mechanism and the genetic structure of viruses." He was born in Berlin, Germany, and died in Pasadena, California. (Courtesy of the Archives, California Institute of Technology.)

BOX 2.1 MAX DELBRÜCK

Max Delbrück was born in Berlin on September 4, 1906. His interest in science was initially directed toward astronomy but then shifted to theoretical physics during his graduate studies at Göttingen University, which was influenced by the breakthrough of quantum mechanics. Delbrück spent three postdoctoral years abroad (1929–1932), in England, Switzerland, and Denmark, where he was exposed to strong cultural and scientific influences including contacts with Wolfgang Pauli and Niels Bohr. Then in 1932 he moved to

Berlin, as assistant to Lise Meither, at the Kaiser Wilhelm Institute for Chemistry. Soon after, Delbrück joined a small group of biologists and physicists who met privately, mainly in the home of Russian geneticist N.V. Timofeeff-Ressovsky. Out of these meetings and their experimental research grew a seminal paper written by Timofeeff-Ressovsky, Zimmer, and Delbrück, which was published in 1935 and is known as the Green Pamphlet. This paper created the background for Schrödinger's book *What Is Life*? (1944), which strongly influenced the initial development of molecular biology.

In 1937 Delbrück's second fellowship from the Rockefeller Foundation supported his move to Caltech, to work in the laboratory of Thomas Morgan, because of its position in *Drosophila* genetics. This allowed Delbrück to escape the dramatic events that followed in Germany. Steadily his interests shifted toward investigation of phages, bacterial viruses. In 1939 Delbrück accepted an instructorship in the physics department at Vanderbilt University in Nashville, Tennessee. There he developed a very fruitful collaboration with Salvador Luria, a young biophysicist who some time later moved to Bloomington, Indiana. Their elegant and very convincing paper on the nature of random mutations, published in 1943, became the basis for the positive Nobel Committee decision in 1969.

In 1947 Delbrück returned to Caltech as a professor of biology where he worked for the next 30 years. In the early 1950s Delbrück began studies of sensory physiology rather than genetics. He also set up the institute for molecular genetics at the University of Cologne and helped to establish the department of biology at the newly founded University of Constance in Germany. The Institut für Genetik der Universität Köln was formally dedicated on June 22, 1962, with Niels Bohr as the principal speaker. His lecture, entitled "Light and Life—Revisited" commented on his original one of 1932, which had been the starting point of Delbrück's interest in biology. It happened to be Bohr's last formal lecture.

Delbrück was a highly influential figure among physicists who moved to biology in the middle of twentieth century. He was honored by numerous awards, including the naming of the influential research organization "Max Delbrück Center for Molecular Medicine (MDC) in Berlin-Buch" not far from the place where he made his first steps toward molecular genetics. Max Delbrück died on March 9, 1981.

Source: http://nobelprize.org/nobel_prizes/medicine/laureates/1969/delbruck-bio.html;http://en.wikipedia.org/wiki/Max_Delbr%C3%BCck

Figure 2.2 Nikolay Timofeeff-Ressovsky (1900–1981), outstanding Russian geneticist and radiobiologist. He was born in Moscow and died in Obninsk, Russia. The signature in Cyrillic on this friendly cartoon drawn by S. Tulkes says: "This is me. N.T-R".

BOX 2.2 NIKOLAY TIMOFEEFF-RESSOVSKY

Nikolay Vladimirovitch Timofeeff-Ressovsky was born in Moscow on September 20, 1900. He began his higher education at a private Moscow university in 1916, which was soon interrupted by the 1917 revolution and the civil war. As a follower of anarchist Peter Kropotkin he joined the Green Army, the third force during the civil war. The two others were the Red and the White Armies. Eventually the Green Army became a part of the Red Army. Barely alive after a life-threatening disease and hunger, he returned to a devastated Moscow in 1920. Despite a lack of formal education and a degree he began teaching and continued his informal education. Soon he met and married Elena Fidler, who was a student of the outstanding biologist N.K. Kol'tsov, director of the Institute of Experimental Biology. She became Timofeeff-Ressovsky's lifelong companion and partner in research.

From 1921 to 1925 Timofeeff-Ressovsky worked at the same institute in the department headed by S.S. Chetverikov, one of the founders of population genetics. The intellectual atmosphere in the institute was very encouraging and young researchers advanced quickly. American geneticist Hermann Muller visited the institute in 1922 and informed colleagues about the latest progress in *Drosophila* research, which was very stimulating for young Timofeeff-Ressovsky.

In 1925 Oskar Vogt, eminent German neurophysiologist and director of the famous Kaizer Wilhelm Institute (KWI) for Brain Research asked Kol'tsov to suggest a promising geneticist for his institute. Thus Timofeeff-Ressovsky was invited to Germany to establish a new genetics laboratory at Vogt's institute. Despite his strong emotional ties to Kol'tsov and to his homeland, Timofeeff-Ressovsky accepted this invitation and moved to Berlin. At that time, he had published just a few papers and was essentially unknown outside of a small circle of Russian biologists. In the following fifteen years between his arrival in Germany and the outbreak of the war, Timofeeff-Ressovsky took a well-deserved position among the leaders of European genetics.

Nikolay and Elena Timofeeff-Ressovsky began their research in Berlin on radiation and experimental population genetics, as well as on microevolution. His excellent scientific results and charismatic personality were taken into consideration when in 1929 he became director of the Department of Experimental Genetics at the KWI, which in 1930 moved to a new facility and was partly financed by a grant from the Rockefeller Foundation. The institute was at that time one of the largest and most modern research facilities of its kind in the world. Together with colleagues and alone Timofeeff-Ressovsky published a number of influential papers. The young Russian geneticist developed close relationships with leading physicists such as N. Bohr, P. Dirac, P. Jordan, and E. Schrödinger, and biologists C. D. Darlington, T. Dobzhansky, Å. Gustafsson, J. Haldane, H. Muller, and N. Vavilov. In 1932 he began collaboration with M. Delbrück and K. Zimmer. The results of this study were published in 1935 and considered to be a major advance in understanding the nature of mutations and gene structure.

(Continued)

BOX 2.2 (CONTINUED)

Timofeeff-Ressovsky stayed in Germany even after Hitler came to power in 1933 and surprisingly he did not have troubles as a USSR citizen. In 1937, Nikolay and Elena Timofeev-Ressovsky were ordered back to Moscow by the Soviet government. At a time when many leading geneticists were prosecuted and even murdered, the return could be suicidal. He rigorously continued research and further advancement of his department. Timofeeff-Ressovsky did not move to the West when WWII ended and met the Red Army as deputy-director of the Institute in May 1945. After the fall of Berlin, Timofeeff-Ressovsky was arrested by NKVD (contemporary acronym for the KGB) but soon after he was released and appointed as the director of the entire institute. The Soviet nuclear project needed a highly experienced radiobiologist such as Timofeeff-Ressovsky. However, in a short while he was secretly re-arrested by different NKVD officers, received a ten year sentence and was incarcerated in the Gulag. There he met future writer and Nobel laureate Alexander Solzhenitsyn, who described this fact in his novel. Due to the harsh conditions and malnutrition Timofeeff-Ressovsky's health rapidly deteriorated. Frédéric Joliot-Curie, a Nobel laureate in chemistry and the leading French scientist, pleaded with the Soviet government that Timofeeff-Ressovsky should be saved and given meaningful work. Eventually, after medical treatment he was sent to a secret institute in the Ural region (Sungul), where prisoners were doing research. In 1955, the Sungul institute was disbanded. Timofeeff-Ressovsky had completed his ten year sentence. His release was an emotional event particularly in Moscow, where a crowd of intellectuals greeted him at a railway station. Timofeeff-Ressovsky organized and became the head of the department of radiobiology at the Institute of Biology in Sverdlovsk (now Ekaterinburg). His famous summer schools contributed to the revival of genetics in the USSR despite governmental prohibition and pressure.

In 1964, Timofeeff-Ressovsky became the head of the department of radiobiology and genetics at the Institute of Medical Radiology in Obninsk. In 1969 Max Delbrück traveled to the USSR and unofficially met with Timofeeff-Ressovsky. Nikolay Timofeev-Ressovsky died in Obninsk on March 28, 1981. In 1991, ten years later, the prosecutor general of the USSR stated that the charge of treason against Timofeeff-Ressovsky in 1946 had no legal basis.

The scientific legacy of Nikolay Timofeeff-Ressovsky is extensive and covers numerous areas of genetics and biology.

Sources: http://www.genetics.org/cgi/content/full/158/3/933; http://en.wikipedia.org/wiki/Nikolay_Timofeeff-Ressovsky

that after some months Delbrück was so deeply interested in quantitative biology, and particularly in genetics, that he stayed in this field permanently" (Zimmer 1966, cited from Ratner 2001).

At the same time, in September of 1932, Hermann Muller (Figure 2.3 and Box 2.3) commenced his sabbatical in Timofeeff-Ressovsky's laboratory, which at that time was one of the best genetics laboratories in Europe. Muller was an outstanding American geneticist and future Nobel laureate, who discovered the mutagenic effect of X-rays and made many other momentous contributions to genetics. The presence of as brilliant and experienced a scientist as Muller in the laboratory was fruitful. He spent a year in the laboratory and also met Niels Bohr and Max Delbrück. In 1933 the political situation in Germany changed dramatically and Muller moved to the USSR where he worked very successfully until Stalin's repressive actions against geneticists began.

Figure 2.3 Hermann Muller (1890–1967), outstanding American geneticist and Nobel laureate (1946) "for the discovery of the production of mutations by means of X-ray irradiation." He was born in New York City and died in Indianapolis. (Courtesy of the Indiana University Archives.)

BOX 2.3 HERMANN JOSEPH MULLER

Hermann Joseph Muller was born in New York City on December 21, 1890. He entered Columbia College at the age of sixteen and became interested in biology. After obtaining a B.A. in 1910 Muller continued his studies at the graduate school. His major interests were in the *Drosophila* genetics work of Thomas Morgan's laboratory, where Muller met undergraduates Alfred Sturtevant and Calvin Bridges, future brilliant geneticists. He formally joined Morgan's group in 1912. Then two years later Muller accepted a position at the W.M. Rice Institute where he promptly completed his Ph.D. and moved to Houston for the beginning of the 1915–1916 academic year. In 1918 Muller proposed an excellent explanation for the famous and then mysterious de Vries's experiments assuming a system of complex translocations, which was a significant contribution to the emerging chromosome theory. Further investigations confirmed his explanation. Muller's work was increasingly focused on mutation rate and lethal mutations. In 1918, Morgan invited Muller to return to Columbia to teach and to expand his experimental program. Soon after, in 1919, Muller made the important discovery of a chromosomal inversion that appeared to suppress crossing-over, which later became instrumental for his mutation studies. The following year he accepted a position at the University of Texas where he worked until 1932 and accomplished his major work. In 1923, he began studying the possible influence of radiation on mutation process. Muller introduced a highly efficient and elegant method for counting lethal mutations and finally after a few years of intensive work demonstrated a clear quantitative connection between radiation and the appearance of lethal mutations. This discovery soon made him a celebrity and won him the Nobel Prize in 1946. Muller made a remarkable contribution to publicizing the danger of radiation exposure in humans and other species.

In 1932 Muller moved to Berlin and spent the following year in one of the most prestigious genetics laboratories of Europe headed by the Russian geneticist Nikolay Timofeeff-Ressovsky. In Germany he met Niels Bohr and Max Delbrück, who had recently become interested in biology. Muller did not want to stay in Germany after Hitler came to power, and moved to the USSR. At the Institute of Genetics initially in Leningrad (Saint Petersburg) and then in Moscow he supervised a large and very successful laboratory of experimental genetics and also commenced work on medical genetics. Stalin's repressions particularly against geneticists made his further work in the USSR impossible and Muller eventually returned to the United States in 1940. Muller's political views and his pacifism did not make his life and career easier, even after becoming Nobel laureate in 1946. He moved from college to college until he finally got tenure at the Indiana University. Muller's research interests were wide. In addition to radiation genetics, which was his central interest, he made an important contribution to population genetics where he formulated the idea of genetic "load." Other interests included the genetics of aging, medical genetics, and eugenics. Prof. Muller retired from Indiana University in 1964. His final appointment just for a year was at the Institute for Advanced Learning in the Medical Sciences, the City of Hope, Duarte, California. Muller made a deep impact in several fields of genetics and was honored by numerous awards and degrees. Hermann J. Muller died on April 5, 1967 in Indianapolis.

Sources: http://en.wikipedia.org/wiki/Hermann_Joseph_Muller; http://nobelprize.org/nobel_prizes/medicine/laureates/1946/muller-bio.html

One of the important results of the collaboration between Timofeeff-Ressovsky, Zimmer, and Delbrück was the publication in 1935 of the classic paper "Über die Natur der Genmutation und der Genstruktur," known also as the Green Pamphlet. This paper provoked Schrödinger's famous book *What Is Life?* first published in 1944. Although the Green Pamphlet did not resolve the question of the molecular nature of genes and mutations, it is commonly considered a very important effort in paving the way for molecular genetics. In this paper an attempt was made to link spontaneous mutations to hypothetical molecular structures and even to possible subatomic interactions. A great intellectual effort indeed!

The term *spontaneous mutation* was introduced by Thomas Morgan, one of the fathers of genetics and Nobel laureate, as early as 1910, when the first such mutation was found in his laboratory. Later in 1927 Muller coined the term *induced mutation*, to describe genetic changes observed in his X-ray experiments. In the Green Pamphlet both types of mutations were treated as representations of the same molecular phenomenon. A connection between quantum uncertainty and the origin of mutations was not directly discussed in the paper but it was "in the air." Schrödinger expressed this idea a decade later in his book *What Is Life?* considering this statement as obvious. Further experimental evidence had shown that intuition had not failed him.

The majority of questions relevant to spontaneous mutations and the molecular nature of genes, however, remained unanswered. Delbrück, motivated by the hope of finding new laws of physics by studying the basic principles of life, took another Rockefeller fellowship and moved to Thomas Morgan's laboratory at the California Institute of Technology (Caltech) in Pasadena in 1937. Then he continued at Vanderbilt University in Nashville, Tennessee. Eventually in 1940 he met biophysicist Salvator Luria, the recent émigré from Italy, and invited him to Nashville. The successful collaboration continued after Luria move to Indiana University in Bloomington. A few years later in 1943 they published a paper on the so-called fluctuation test, which eventually led to the Nobel Prize in 1969 together with Alfred D. Hershey. They used a new model system of bacteria and phages (bacterial viruses) and formed the so-called Phage Group. One of the objectives was to figure out whether bacterial resistance was the result of some action of phages on bacterial cells or the resistance was caused by random mutations. The story goes that Luria's idea was inspired by watching people playing slot machines at a country club. The payoff from a slot machine varied from a few coins, which was rather common, to a large amount, a rare occurrence. He thought a similar pattern could be expected in the development of phage resistance in bacteria. This idea led to the development of the fluctuation test, which was elegantly supported by Delbrück's mathematical analysis.

Luria and Delbrück tested two hypotheses. The first was the mutation hypothesis, which assumed that rare random mutations occur at a constant rate and do not depend on the presence of phages in the bacterial culture. The second was the acquired immunity hypothesis, that each cell in the culture may with a small probability survive the phage attack and this is somehow related to the interaction between phages and bacterial cells. The test clearly indicated that only the mutation hypothesis fit the experimental data. Thus, bacterial resistance to phages was caused by spontaneous mutations, which occur rarely and randomly in bacterial cells regardless of the presence of phages. In some way this result confirmed conclusions earlier made in the Green Pamphlet; however, it was done in such a rigorous way that the question was finally put to rest. The Darwinian paradigm that an adaptive response is caused by spontaneous, random mutations was strongly supported by these irrefutable results.

Still the problem of the chemical nature of genes remained unresolved. The breakthrough came with the publication of James Watson and Francis Crick's paper on DNA structure on April 25, 1953, in the British journal *Nature*. James Watson was the most famous student of Luria, the junior member of the Phage Group and a friend of Delbrück. Interestingly Delbrück was the first to learn about this remarkable discovery prior to its publication in *Nature* through a letter from Watson on March 12, 1953. He was also the first to invite Watson to deliver a lecture on the double helix. Crick (1988) wrote: "Mainly due to Max Delbrück, copies of the initial three papers were distributed to all those attending the 1953 Cold Spring Harbor Symposium, and Watson's talk on DNA was added to the program."

The generation of young physicists who entered biology in the 1940s and 1950s was deeply inspired by the problems discussed briefly in this section. Among them were Francis Crick, Seymour Benzer, and many other bright researchers who made an exceptional contribution to molecular genetics. During the next three decades many well kept secrets of the molecular organization of life were discovered. It became clear that changes in DNA are the cause of mutations. The connection between events on a subatomic level and phenotypes of living creatures was firmly established. This is discussed briefly in the following section. A conclusion can be drawn that quantum uncertainty observed on the level of elementary particles has an impact on macroscopic events in living matter.

Mutations and repair

Types of mutations

The term *mutation* commonly used in genetics was introduced in the late nineteenth century by the famous Dutch researcher Hugo De Vries (1848–1935), one of the first geneticists. It took nearly seven decades of intensive investigations before the nature of mutations became clear. Modern understanding of the molecular mechanisms of mutations, mutation rate, and factors causing or preventing mutations has grown rapidly since the 1950s and is widely used in everyday life.

It is common knowledge that genes are sections of very lengthy DNA molecules. DNA is a double helix, that is, a molecule with two complementary chains or strands built from nucleotides (Figure 2.4). Each nucleotide consists of a deoxyribose sugar, a phosphate group, and one of four nitrogenous bases. That is why there are four types of nucleotides: A (adenine), C (cytosine), G (guanine), and T (thymine). The nucleotides from the opposite DNA strands form complementary pairs according to the rule: A and T; G and C. A·T and G·C pairs are connected by so-called hydrogen bonds; two such bonds for A·T and three for G·C pairs. Replacement of one nucleotide by another in DNA constitutes a mutation, usually called substitution, a very frequent type of mutation. There are many other types of mutations, including deletions and insertions of small or large sections of a gene or a chromosome, which may carry numerous genes. Another commonly used term for a substitution is a *point mutation,* so called because only one pair of complementary nucleotides is affected, which represents a tiny dot on the lengthy DNA molecules.

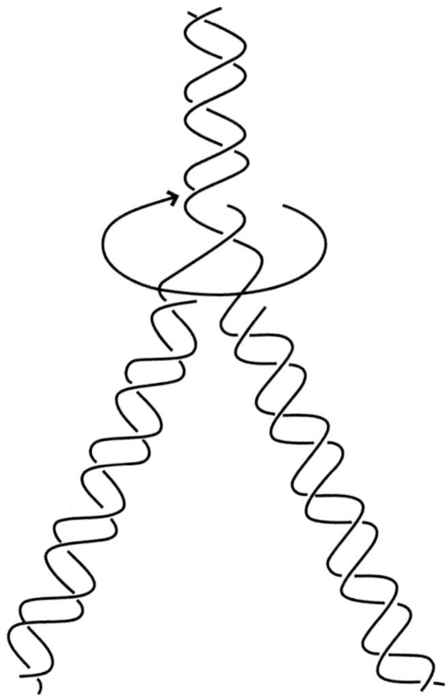

Figure 2.4 DNA double helix and replication. DNA is a linear molecule with two strands built from four nucleotides (A, C, G, T). Nucleotides from the complementary strands form pairs according to the rule: A matches T and G matches C. During the replication the double helix is unwound and each strand acts as a template for the newly synthesized complementary strand.

Keto-enol transitions and quantum uncertainty

The same organic molecule may exist in more than one molecular form. There are several processes causing such transitions. Tautomerism is one of them and changes a molecule from the keto to the enol form and back (Figure 2.5). The mechanism of keto-enol transition was investigated by several outstanding chemists and is well understood. This involves repositioning of bonding electrons and movement of a proton; otherwise the molecule does not change. Keto-enol equilibrium is usually biased in favor of the keto form. Nucleotides in a DNA molecule also undergo spontaneous tautomeric shifts. As an example, two tautomers of guanine are shown in Figure 2.6. The difference between these alternative forms is in the distribution of electron density in the molecules and shift of a proton. The keto form is the regular and stable one; the enol form is a much rarer variant. This fact was not known as yet when Watson and Crick were building models of DNA in 1951–1952 and the lack of this knowledge complicated their effort. As soon as the double helix model was built they

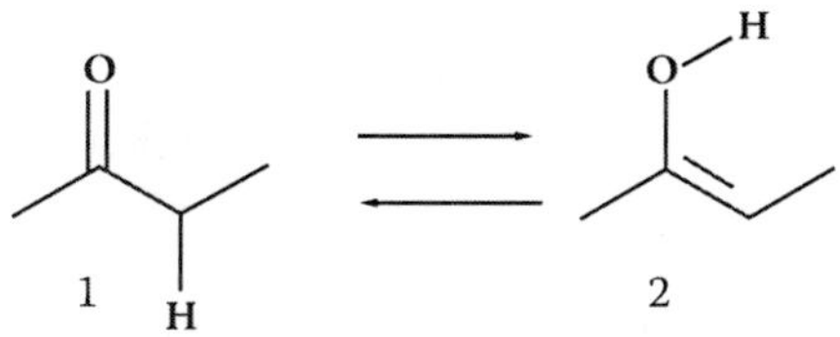

Figure 2.5 Keto-enol tautomerism. (1) is the keto and regular form; (2) is the enol and rare form. The transition from one form to another involves the movement of a proton and the repositioning of bonding electrons. Only oxygen (O) and hydrogen (H) atoms are shown. Single line represents one covalent bond.

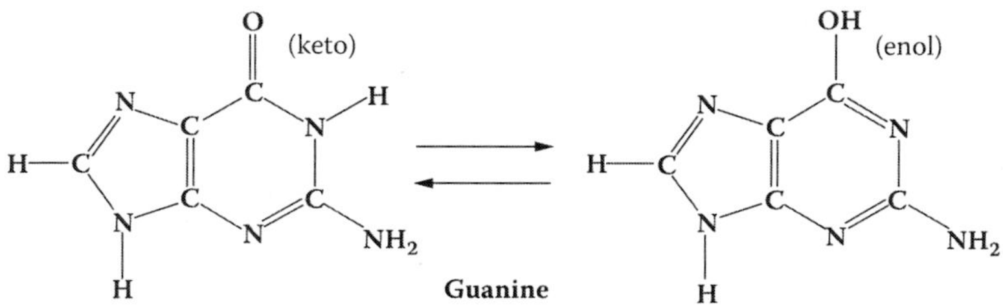

Figure 2.6 Guanine can be found in either the regular keto form or the rare enol form and transitions between the forms occur from time to time.

immediately saw a possibility that a tautomeric shift could be a potential cause of mutations. Tautomeric transitions being instantaneous and reversible usually do not affect structure of DNA molecule unless such transitions happen during DNA replication.

Another possible quantum effect, which might lead to tautomeric shift, is proton tunnelling when a proton can move through the energy barrier. This idea was first suggested by Löwdin in 1965. Although there have been several recent attempts to verify this hypothesis, the final evidence is still missing. It seems that nobody has ruled out this possibility but it is nevertheless unclear how significant such quantum contribution could be.

DNA replication is a unique and absolutely essential process. Life, as we know it, is not possible without replication. The two-stranded complementary structure of the DNA molecule is the basis for replication. While the process is very complex and served by many enzymes, it contains two critically important steps (Figure 2.4). The first is unwinding the double helix, which opens opportunity for the complementary synthesis. The second is synthesis itself when free nucleotides are assembled on each of two DNA strands acting as templates. Nucleotides in the newly synthesized strand match to nucleotides from the template according to the rule: A matches T and G matches C. The fidelity of DNA replication is high but mistakes are inevitable. Several proofreading and repair mechanisms constantly correct or repair replication errors.

Now let us consider one of many possible causes of such errors (Figure 2.7). As already mentioned, a tautomeric shift may occur during DNA replication, for example, from the keto form of guanine to the enol form (Figure 2.6). The enol form is more prone to mispairing than the common keto form. As a result a G·T mispair could appear. Unless this mispair escapes correction by proofreading or repair systems, a new mutation will occur (Figure 2.7). As the figure shows after the replication one of the daughter DNA molecules carries the A·T pair instead of parental G·C, that is, a substitution took place.

Nucleotides differ in their ability to be involved in tautomeric shifts. For instance, the frequency of tautomeric shifts of cytosine to its enol form is considerably lower than of guanine and hence the contribution of cytosine to the spontaneous point mutations of G·C → A·T type is insignificant (Podolyan, Gorb, and Leszczynski 2003). It means that the term random in this context does not imply equal probability for possible different mutation events; it rather stresses the uncertainty and probabilistic nature of the process.

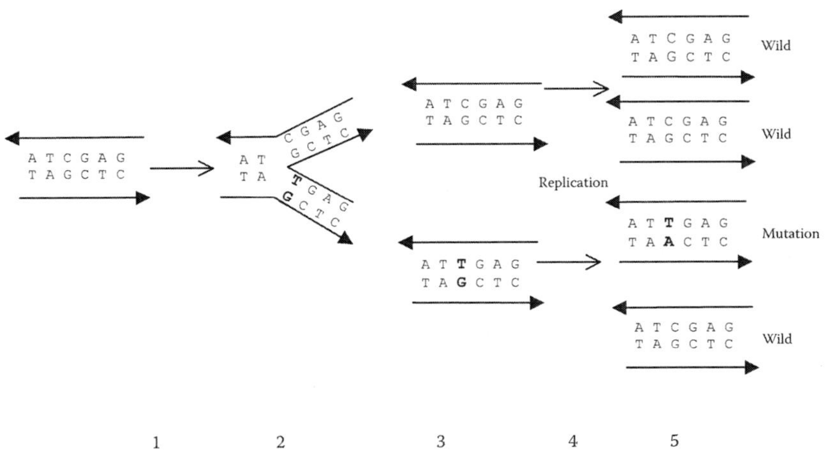

Figure 2.7 Mutation in DNA caused by the tautomeric shift during replication. 1. DNA replication. 2. Infrequent and temporary transition from common keto form of guanine to rare enol form may lead to a breach of the complimentarily rule. Instead of canonic G·C pair existing in parental DNA in this position, a mispair G·T may occur. 3. If this replication mistake is not removed or repaired, a mismatch or in other words pre-mutation would be fixed in one of two daughter DNA molecules. 4. Next round of DNA replication. 5. Replication of the DNA molecule with the mismatch will lead to mutation. The original nucleotide pair G·C in this position is substituted by the mutant pair A·T.

Thus the tautomeric shifts could lead to DNA mutations. Is there any particular reason for a tautomeric shift? This is the question without a definitive answer because the subatomic world is governed by the uncertainty principle. Causality is not a part of the quantum rules and electron movements that eventually lead to the tautomeric shift. Randomness at this level seems to be irreducible and Heisenberg's uncertainty principle precludes exact knowledge of the position and momentum of an electron or other elementary particles. A probabilistic description of such phenomena is the only available option. Similar views are shared by other researchers, including McFadden (2001), who develops this idea in the book *Quantum Biology*. The quantum chemical calculations estimating frequency of the tautomeric shifts from the standard Watson–Crick pair to a mispair have been made and show a reasonable correspondence to the spontaneous mutation rate (Kryachko and Sabin 2003). Transformation of a quantum fluctuation into a mutation is the essence of the process. In philosophical terms it can be interpreted as generation of certainty from randomness. This process occurs numerous times per hour in each living cell and only robust proofreading and repair systems can reduce the quantity of new mutations.

Mutations could also be promoted at the level above subatomic. Thermal or, in other terms, Brownian motion, is also a potential source of substitutions. DNA polymerase, the enzyme responsible for DNA replication, being a very large protein molecule, is a subject of Brownian motion. It looks likely that thermal motion causes some infidelity of DNA polymerase, which in turn may lead to substitutions.

Each and every mutation is unpredictable. Questions such as when or why the next mutation will occur do not have an answer and, assuming a quantum nature, at least some of them never will. As mentioned earlier, more than sixty years ago Schrödinger (1944), even without knowledge of DNA structure and the molecular nature of mutations, drew a similar conclusion that quantum fluctuations might be the cause of some mutations. This conclusion was rather based on general principles. While the founders of quantum mechanics clearly outlined the universality of quantum laws, up until today application of these principles to living organisms has not been the prevailing trend. A gap between subatomic events and biological processes looks too great for the direct association of quantum phenomena and processes of life. Biologists are usually preoccupied with problems that might be resolved by experimental or theoretical methods. However, if one assumes that quantum events lead to mutations, then uncertainty and unpredictability of this basic biological process is inevitable.

Such a view certainly does not diminish the great value of fundamental genetic concepts and it is not agnostic. Nevertheless, it does indicate the principle of the limits of knowledge and builds a bridge between two very remote levels of organization: the subatomic and macroscopic levels of living creatures. Statistical description of mutation processes in well-studied species is good enough for predictive purposes, which may take into account radiation level, concentration of chemical mutagens, efficiency of repair mechanisms, and many other natural factors.

Induced mutations and DNA repair

The difference between induced and spontaneous mutations is arbitrary and can be explained by a lack of direct knowledge about the cause of each spontaneous mutation. The best way to learn about potential causes of spontaneous mutations is studying induced mutations when the nature of a mutagenic factor is known and the intensity is controlled. Tautomerization is not the only source of new mutations; there are many others. Let us take as an example ultraviolet (UV) light induced mutations. UV light causes the appearance of unusual chemical bonds between neighboring thymines located in the same DNA strand. If such abnormal T=T structures are not reversed to a normal condition or repaired, it will block normal DNA replication (Setlow and Carrier 1964). During replication of the UV damaged segment, DNA polymerase is not able to recognize the content of this segment and either stop further synthesis or randomly insert nucleotides in the growing strand. If proofreading and repair mechanisms do not reverse the DNA structure back to normal, mutations are inevitable. UV is a very common radiation and life was dealing with it for billions of years. Unless this problem is effectively controlled, life is under a serious threat. That is why there are several defence mechanisms dramatically reducing the risk of new UV generated mutations.

The efficient way of reversing the T=T structure to a normal one, when the neighboring thymines are no longer linked by chemical bonds, is so-called photoreactivation. A special enzyme called photolyase binds to the thymine dimer and splits it, thus reversing the damage. This enzyme needs energy to act, which comes from visible light. Alternatively another group of enzymes cut out the entire damaged segment in one DNA strand and replaces it with the correct sequence using the second undamaged strand as a template. If a cell is overwhelmed by numerous damaged segments there is a system of emergency response called SOS repair. This system helps DNA polymerase to bypass the damaged segments but the cost of such emergency help is a lack of precision. As a result SOS supported replication may generate new mutations (Woodgate 2001).

The photons initiating the entire process are elementary particles with quantum mechanical characteristics and their randomness is intrinsic and unavoidable. The UV photons randomly hit nucleotides in DNA molecules and initiate the ultrafast (~1 picosecond) mutagenic photolesion points in the nucleotide bases that are properly oriented at the instant of light absorption (Schreier et al. 2007). Another layer of randomness might be added to the process by the SOS system, which from time to time incorporates erroneous nucleotides in the newly synthesized DNA strands and also causes additions and deletions of nucleotides.

The repair systems play an exceptionally important role in controlling the intensity of the mutation process. The first knowledge of such systems was obtained in the late 1940s by Albert Kelner, then at Cold Spring Harbor Laboratory, and Renato Dulbecco (1975 Nobel Prize laureate) at Salvador Luria's laboratory. Since then a large number of different and sometimes very sophisticated repair systems have been discovered. It became clear that life without efficient repair systems is nearly impossible. The most impressive example of this is xeroderma pigmentosum, an autosomal recessive genetic disorder observed in humans in which the ability to repair DNA damage caused by UV light is compromised. Individuals with this condition develop numerous skin cancers at a young age. In order to avoid the consequences of the constant bombardment of genomes by newly arisen mutations, each human cell repairs several thousand mutations caused by depurination, and many hundreds of mutations generated by deamination, methylation, and oxidation every day (Lindahl 2000). Fortunately the majority of normal individuals are shielded from ever arising mutations by general and specific repair systems.

There are a few more types of DNA damage resulting from diverse molecular interactions. All of them most likely are random regardless of their special features. Either quantum or stochastic effects are typical sources of these mutations and their differentiation is hardly possible. Due to the high efficiency of repair systems DNA replication is remarkably accurate. Wrong nucleotides can be found in newly synthesized DNA approximately every 10^9 to 10^{10} nucleotides. Thus, the probability of a mutation event per nucleotide is not high at all. However, the number of nucleotides, for instance in the human nuclear genome, is huge ($\sim 3 \times 10^9$). This roughly means that one new mutation event occurs during replication of a single cell.

How do random molecular events like mutations become facts of life?

In the tautomeric scenario that we have considered, there are three critically important steps (Figure 2.7). The first one is a shift from the keto to the enol form of guanine and the possibility of a mismatch with a complementary nucleotide in the newly synthesized DNA strand. This first step can be characterized as necessary but not sufficient for producing mutation. Exactly the same tautomeric shift occurring prior to or after DNA replication does not lead to a mutation because of the very quick transition from the enol back to the keto form. The second step is synthesis of the complementary DNA strand during the same replication round which entails an abnormal match of the enol form of guanine (G) and normal thymine (T). In this event a potentially long lasting change in the DNA molecule might occur, which can be characterized as sufficient for the mutation process. Certainly, this is not the final verdict and the repair mechanisms will most typically redo such potential change. Nevertheless this is the step during which all DNA replication related mutations are generated. In other words the crucial imprint on the newly synthesized DNA strands is made. Until the next replication round the G·T mispair is vulnerable and, strictly speaking, is not yet a proper mutation but rather a premutation, as there is a high likelihood of correction when T can be replaced by C. However, if such correction does not occur, the situation changes completely after the following replication round. This is the third and the final step of the process, when the newly appeared nucleotide pair A·T became "legitimate" and chemically indistinguishable from any other pair. As soon as the substitution of a G·C pair by an A·T pair is accomplished, this mutation is finally "fixed" in the structure of a DNA molecule. At this point quantum and chemical aspects of the mutation process are behind. However, it is only the beginning of assessment of a new mutation by the forces of natural selection as well as it becomes a subject of random population events described in Chapter 6.

Somatic and germ cell mutations

The fate of a mutation in multicellular species like animals, plants, or fungi very much depends on the cellular type where it occurs. If this is a somatic cell, such as liver, brain, or muscle, then a mutation obviously does not have a chance to be passed to the next generation. If this is a relatively neutral mutation it might either spread over or alternatively be lost as a result of random events taking place in populations of dividing cells. Certainly the ongoing DNA replication is the vehicle for spreading this mutation. Those mutations that have detrimental effects on cellular processes will be likely wiped out due to their death or low reproductive capacity. However, from time to time some mutant somatic cells obtain a significant reproductive advantage, like in the case of malignant cells. Then the survival of such cells might depend on the survival of an individual.

Mutations may also occur in germ cells, like sperm or eggs or their predecessors. Such mutations could indeed be passed to the next generations of individuals and they are absolutely essential for evolution. The germ mutations may also be either spread or lost in populations, a complex process which will be considered later in the book. In both somatic and germ mutations transformation of a rare quantum event into a major phenomenon of life is based on DNA replication. No other molecule, except

RNA, is capable of replication. This incredibly powerful force can produce huge quantities of cells and individuals carrying a particular mutation in a relatively short time, thus effectively copying the earlier "fixed" quantum effect. Such potential increase in numbers could be exponential and always based on DNA replication. This is truly the distinctive feature of living matter. From the very earliest stages of biological evolution quantum novelties were conserved in the structures of replicating molecules. Timofeeff-Ressovsky called this fundamental principle of life "convariant replication," in other words replication of DNA molecules with constantly occurring mutations. The differential amplification of spontaneous mutations initiated by quantum events and governed by natural selection and random population genetics events is the key feature of organic evolution. A similar process is not possible in nonliving matter because the replication mechanism does not exist there and this is a very important distinction. Mutations are the basic "bricks" of variability but there are other very potent sources of variation observed in natural populations, which are considered in the following chapters.

Quantum uncertainty and unpredictability of life

As we saw in this chapter quantum effects are directly related to mutations. Therefore, indeterminism originating in the subatomic world can impact genetic variability. Despite a colossal gap between the subatomic world and organismal level the connections between them are real. Of course this does not mean that description of life could be reduced to physical or chemical processes only. Such reductionism was criticized in the past and probably will be in the future. Here we attempt to add to this long lasting debate by emphasizing that physical description of biological processes is not only impractical in many cases but impossible in principle due to ever-present layers of randomness. On the other hand the notion that physical laws are universal and relevant to all forms of matter is certainly valid.

Still this is quite different from the core idea of this chapter. There is no doubt that the principles of quantum mechanics are applicable to elementary particles and atoms composing living creatures. The main message, however, is that some quantum events directly affect the entirely different levels of organization, namely macroscopic living objects. One could probably say that such knowledge can be ignored in biological research without measurable consequences and this might be correct. Nonetheless, it could be very useful for biological thinking as well as for the philosophy of science because it shows the limits of knowledge as well as the interactions between distinct and separate levels of matter.

J. Monod, an outstanding molecular biologist and Nobel laureate (1965), made an important contribution to understanding of the role of chance in nature in his renowned book *Chance and Necessity* (1971).

During the last decade or so the question about the role of quantum indeterminism in biological processes was debated again. Some researchers (Brandon and Carson 1996; Stamos 2001) supported the idea that quantum uncertainty inevitably contributes to the basic biological phenomena, including the mutation process. Their opponents (Graves, Horan, and Rosenberg 1999) argued that the processes of evolutionary biology are fundamentally deterministic and the only way they are described is statistical. A decisive proof in a philosophical debate is a rare commodity and this discussion is not an exception. Nonetheless, the totality of the data as well as steady progress in the development of a probabilistic view of living matter should eventually sway such debates in favor of indeterministic views.

Other quantum phenomena and life

Are there other opportunities for quantum phenomena in biological systems? This question was considered in several publications (Davies 2004; McFadden 2001) and, although the authors believe that stronger evidence is still required, such effects seem possible. For instance quantum tunnelling seems essential for the exceptional ability of enzymes to catalyze biochemical reactions. How this might affect intrinsic randomness of biological systems remains to be understood. Unfortunately this interesting and complex topic is beyond the scope of this chapter and the book.

In an unexpected way quantum computer calculations might also be related to the basic life processes. The theoretical work by Patel (2001) provides an insight. He considered replication of DNA and synthesis of proteins from the viewpoint of quantum database search. Patel claimed that "identification of a base-pairing with a quantum query gives a natural explanation of why living organisms have 4 nucleotide bases and 20 amino acids. It is amazing that these numbers arise as solutions to an optimisation problem." He also proposed that "enzymes play a crucial role in maintaining quantum coherence of the process" and suggested experimental tests. Further investigations in this direction hopefully will allow verification of the claims.

Summary

The majority of mutations originate at the DNA level. This is particularly true for nucleotide substitutions, the most common type of mutations. From the first attempts at understanding the molecular nature of mutations, through the proof of their spontaneous nature, to the modern quantum chemistry description of substitutions, we explored some ideas and facts relevant to the mutation process.

The quantum chemistry estimates regarding stability of the premutational A·T mispair fit well into the known range of mutation frequencies per base pair, thus confirming that bottom up and top down approaches do not contradict each other. The major conclusion of this chapter is that some types of new mutations are initiated by the quantum events and therefore timing and location of the next mutation cannot be predicted in principle. Here is the point where the quantum uncertainty principle is fully applicable to genetics. While this fact has been known for some time and it does not change the long standing view on spontaneous mutations, it is important to emphasize the existence of absolute limits of scientific understanding. This chapter also explains how quantum events become stable facts of life and why DNA replication is the crucial process in transforming volatile quantum shifts into the long term changes of DNA structure.

One could question the conclusions made in this chapter on the grounds that modern science has already accumulated a great deal of knowledge about the mutation process. Indeed, in many instances there is comprehensive data explaining how new mutations come into being. However, the answer to the question why does this spontaneous mutation occur is problematic as is prediction of when and where the next mutation will occur. It is true that the probability of new mutations can be dramatically increased by mutagenic agents. Similarly timing and location of new mutations can be narrowed by modifications of target DNA and the mutagenic agents. Nevertheless, the probabilistic description of the mutation process cannot be replaced by a deterministic one, and this is the basic argument in the discussion about the limits of scientific advancement. A similar logical reference is applicable to other genetic phenomena considered in the following chapters. The notion of the limits of knowledge expressed here is not agnostic, it is rather realistic. In a sufficiently large set of data the difference between a deterministic and a probabilistic description of events becomes small. Even an accurate general prediction is possible but for a relatively short time. Long term forecasts of biological processes are not accurate and this is one of the reasons why life is so unpredictable.

References

Brandon, R.N., and S. Carson. 1996. The indeterministic character of evolutionary theory: No "hidden variable" proof but no room for indeterminism either. *Philosophy of Science* 63:315–337.

Crick, F. 1988. What mad pursuit. A personal view of scientific discovery. New York: Basic Books.

Davies, P.C.W. 2004. Does quantum mechanics play a non-trivial role in life? *BioSystems* 78:69–79.

Graves, L., B.L. Horan, and A. Rosenberg. 1999. Is indeterminism the source of the statistical character of evolutionary theory? *Philosophy of Science* 66:140–157.

Klug, W.S., and M.R. Cummings. 1997. *Concepts of Genetics*, 5th ed. Upper Saddle River, NJ: Prentice-Hall.

Kryachko, E.S., and J.R. Sabin. 2003. Quantum chemical study of the hydrogen-bonded patterns in AT base pair of DNA: Origins of tautomeric mispairs, base flipping, and Watson-Crick -> Hoogsteen conversion. *International Journal of Quantum Chemistry* 91:695–710.

Lindahl, T. 2000. Suppression of spontaneous mutagenesis in human cells by DNA base excision-repair. *Mutation Research* 462:129–135.

Luria, S.E., and M. Delbrück. 1943. Mutations of bacteria from virus sensitivity to virus resistance. *Genetics* 28:491–511.

McFadden, J. 2001. *Quantum biology*. New York: Norton.

Monod, J. 1971. *Chance and necessity: An essay on the natural philosophy of modern biology*. New York: Alfred A. Knopf.

Patel, A. 2001. Quantum algorithms and the genetic code. *PRAMANA Journal of Physics* 56:367–381.

Podolyan,Y., L. Gorb, and J. Leszczynski. 2003. *Ab initio* study of the prototropic tautomerism of cytosine and guanine and their contribution to spontaneous point mutations. *International Journal of Molecular Science* 4: 410–421.

Ratner, V.A. 2001. Nikolay Vladimirovich Timofeeff-Ressovsky (1900–1981): Twin of the century of genetics. *Genetics* 158:933–939.

Schreier, W.J., T.E. Schrader, F.O. Koller, P. Gilch, C.E. Crespo-Hernández, V.N. Swaminathan, T. Carell, W. Zinth, and B. Kohler. 2007. Thymine dimerization in DNA is an ultrafast photoreaction. *Science* 315:625–629.

Schrödinger, E. 1944. *What Is Life?* Cambridge, UK: Cambridge University Press.

Setlow, R.B., and W.L. Carrier. 1964. The disappearance of thymine dimmers from DNA: An error-correcting mechanism. *Proceedings of the National Academy of Sciences USA* 51:226–231.

Stamos, D.N. 2001. Quantum indeterminism and evolutionary biology. *Philosophy of Science* 68:164–184.

Timofeeff-Ressovsky, N.W., K.G. Zimmer, and M. Delbrück. 1935. Über die Natur der Genmutation und der Genstruktur. *Nachrichten von der Gessellschaft der Wissenschaften zu Göttingen. Biologie. Neue Folge* 1:189–245.

Watson, J.D., and F.H.C. Crick. 1953. A structure for deoxyribose nucleic acid. *Nature* 171:737–738.

Woodgate, R. 2001. Evolution of the two-step model for UV-mutagenesis. *Mutation Research* 25 (485):83–92.

Zimmer, K.G. 1966. The target theory. In *Phages and the origins of molecular biology*, ed. J. Cairns, G.S. Stent, and J.D. Watson, 33–42. Cold Spring Harbor, NY: Cold Spring Harbor Laboratory Press.

chapter three

Recombination and randomness

> The combination rule is confined by the strange phenomenon that Morgan calls crossing-over or the exchange of genes, which he imagines as a real exchange of parts between the chromosomes. … A German scientist has appropriately compared this to the astronomical calculation of celestial bodies still unseen but later on found by the tube—but he adds: Morgan's predictions exceed this by far, because they mean something principally new, something that has not been observed before.
>
> **Presentation speech by the Nobel Committee**
> (1933)

What is recombination?

In this chapter we shall discuss very different biological processes which have at least one common feature: recombination of DNA molecules. Two DNA molecules, if broken, can rejoin. It serves at least two purposes: repair and production of novel DNA molecules which might carry new genetic information. This is the essence of recombination, an ubiquitous and ancient genetic process. There are several types of recombination and only some of them will be considered briefly in this chapter. The major differences in recombination are between prokaryotes (i.e. bacteria), which have simple cells without a nucleus and a single primitive chromosome, and eukaryotes with highly structured cells and complex chromosomes (i.e. plants and animals). Chromosome recombination in eukaryotes can occur either in mitosis or meiosis.

Mitosis is the typical division of eukaryotic somatic cell. It includes a number of consecutive steps, two of which are of particular importance. One of them is replication of chromosomes, which formally is not a part of the mitotic process itself, but absolutely necessary for mitosis to proceed (Figure 3.1). The second step is segregation of replicated chromosomes, called sister chromatids, during which they move to opposite poles. The daughter cells produced by mitotic division normally have the same number of chromosomes as their parental cells and their genetic makeup is

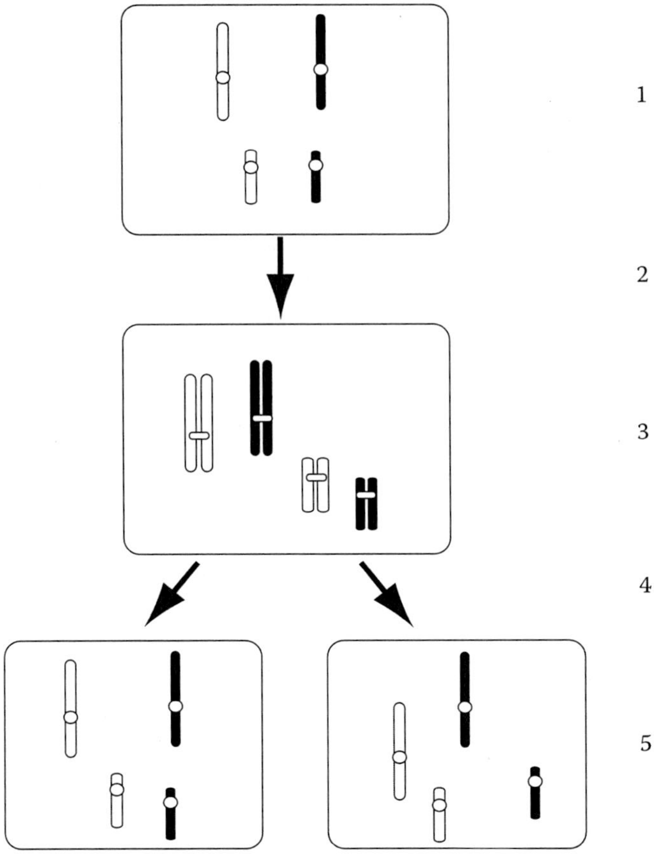

Figure 3.1 Simplified schematic presentation of mitosis. (1) Chromosomes prior to replication (only two pairs are shown; paternal and maternal chromosomes of each pair are painted differently). (2) Replication of chromosomes; nearly identical copies (sister chromatids) of each parental chromosome are made. (3) Chromosomes prior to mitotic division. (4) The mitotic division, sister chromatids are segregated. (5) Two daughter cells with the same set of chromosomes as the parental cell.

almost identical to the parental in the majority of cell types. However, there are important exceptions found in some immune cells. As discussed later in this chapter, a great deal of diversity is generated by recombination events in immune cells.

Meiosis is the major generator of recombination events that occur in germ cells. This is a more complex process than mitosis. As with mitosis it also is preceded by chromosome replication. Then the first meiotic division follows that halves the regular (usually diploid) chromosome number and this is the essence of meiosis (Figure 3.2). Prior to the first

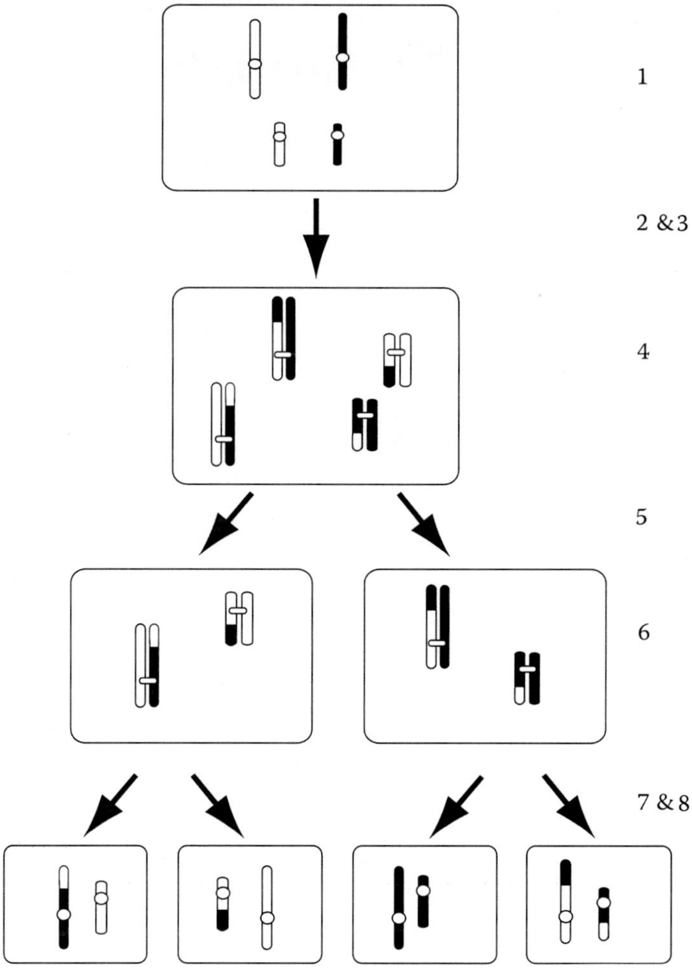

Figure 3.2 Simplified schematic presentation of meiosis, with only some possible outcomes shown. (1) Chromosomes prior to replication (only two pairs are shown; paternal and maternal chromosomes of each pair are painted differently). (2) Replication of chromosomes; nearly identical copies (sister chromatids) of each parental chromosome are made. (3) Homologous chromosomes are involved in crossing-over events and exchange parts at the early stages of meiosis. (4) Chromosomes prior to the first meiotic division. (5) The first meiotic division; homologous chromosomes are segregated. (6) Chromosomes prior to the second meiotic division. (7) The second meiotic division; sister chromatids are segregated. (8) Haploid set of chromosomes in germ cells; each homologous chromosome represented by a single copy.

meiotic division homologous chromosomes are aligned and involved in recombination exchanges, called crossing-over (Figure 3.3). The final step of meiosis is similar to mitotic division.

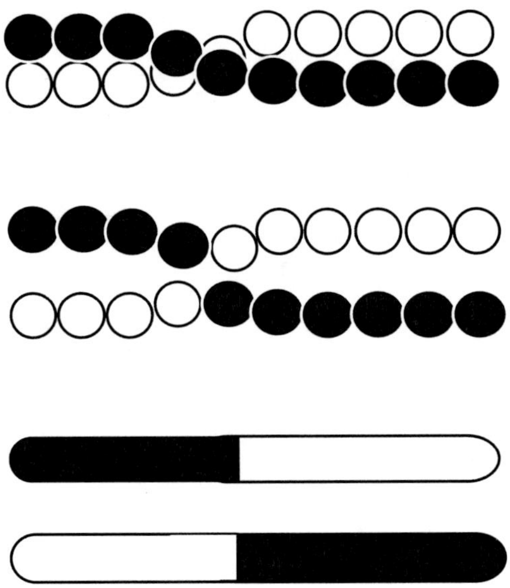

Figure 3.3 Crossing over of chromosomes during early meiosis. Chromosomes undergoing recombination are involved in structures resembling the Greek letter χ and are called chiasmata (plural of chiasma). (Redrawn from Morgan, T. 1916. *A Critique of the Theory of Evolution.* http://en.wikipedia.org/wiki/File:Morgan_crossover_1.jpg.)

Formation of chiasmata (plural of chiasma), χ shaped chromosomes, is the morphological evidence of successful recombination. A connection between microscopically observed chiasmata and genetically deduced crossing-over events started to emerge in the late 1920s due to the efforts of Cyril Darlington, a leading English cytogeneticist. The final experimental proof was presented in 1931 by American researchers Harriet Creighton and Barbara McClintock (Nobel laureate, 1983, for the discovery of mobile genetic elements).

Crossing-over

Each chromosome carries many genes and meiotic exchanges between homologous chromosomes may generate new combinations of alleles. This fundamentally important idea was not suggested until 1910 when Thomas Morgan (Nobel laureate, 1933) (Figure 3.4 and Box 3.1), the founder of the chromosome theory of heredity, conceptualized the results of breakthrough experiments in his *Drosophila* laboratory at Columbia University. The cytological proof of crossing-over was still two decades away and Morgan had to deduce crossing-over using the results of his breeding experiments and excellent imagination. Morgan's deduction

Figure 3.4 Thomas Hunt Morgan (1866–1945), outstanding American geneticist and embryologist, Nobel laureate (1933) "for his discoveries concerning the role played by the chromosome in heredity." He was born in Lexington, Kentucky, and died in Pasadena, California. (Courtesy of the Archives, California Institute of Technology.)

was wonderful. He realized that the recombination events should be a result of chromosome exchanges and he proposed hypothetical χ shaped structures, which should be directly related to recombination events (Figure 3.3). Morgan thought that the frequency of recombination events between genes located on the same chromosome may differ depending on the distance between them. Closely located genes are rarely involved in recombination, while recombination between remotely located genes is very likely. In 1911 Morgan asked Alfred Sturtevant (Figure 3.5 and Box 3.2), at the time his undergraduate student, to make some sense of the crossing-over data. The following is Sturtevant's recollection of these events: "In conversation with Morgan, I suddenly realized that the variation in strength of linkage, already attributed by Morgan to differences

BOX 3.1 T.H. MORGAN

Thomas Morgan was born in Lexington, Kentucky, in 1866. He graduated from the State College of Kentucky (now the University of Kentucky) in 1886 and began graduate studies in zoology at the recently founded Johns Hopkins University. Morgan graduated with a Ph.D. from Johns Hopkins in 1891. In the same year he was appointed associate professor at Bryn Mawr College and three years later he was granted a twelve month absence to conduct research in the laboratories of Stazione Zoologica in Naples, which was a prominent research center in the late nineteenth and early twentieth centuries. There he was exposed to new trends in experimental biology and shifted his interests from traditional morphology to experimental embryology. At the time a significant discussion was going on over the cause of embryonic development. Morgan favored interactions between the protoplasm, the nucleus of the egg, and the environment as the major causes. Soon after returning to Bryn Mawr in 1895 he was promoted to full professor. During the following decade his research was concentrated on regeneration and sex determination.

In 1904 E.B. Wilson, an outstanding biologist, invited Morgan to join him at Columbia University. Morgan took a professorship in experimental zoology and focused his research on the mechanisms of heredity and evolution. Not without a struggle he steadily accepted Mendel's laws. From 1908 Morgan started working on the fruit fly *Drosophila melanogaster*, and encouraged students to do so as well. In 1909, a series of heritable mutants appeared, some of which displayed Mendelian inheritance. A year later Morgan noticed a white-eyed mutant male among the red-eyed flies and using this model he explained the inheritance of sex-linked traits. This discovery was the first step toward the chromosome theory of heredity. Investigation of other mutations and their mode of inheritance led Morgan to the idea of genetic linkage and crossing-over. This was a real breakthrough. Morgan had several exceptional students, including Alfred Sturtevant, Calvin Bridges, and Hermann J. Muller, who greatly advanced the research progress in the famous Fly Lab. Sturtevant developed the first genetic map in 1911 and publication of the work came two years later.

In 1915 Morgan, Sturtevant, Calvin Bridges, and Hermann Muller wrote the seminal book *The Mechanism of Mendelian Heredity*, which became the milestone in development of genetics and the chromosome theory. British geneticist C.H. Waddington noted that "Morgan's theory of the chromosome represents a great leap of imagination comparable with Galileo or Newton." The ideas born in the Fly Lab spread over the world and were generally accepted and confirmed.

In 1928 Morgan became head of the division of biology at the California Institute of Technology. An incredible group of researchers invited by Morgan worked in the division, including Calvin Bridges, Alfred Sturtevant, Jack Shultz, Albert Tyler, Theodosius Dobzhansky, George Beadle, Boris Ephrussi, Edward L. Tatum, Linus Pauling, Frits Went, and Sidney W. Fox.

In 1933 Morgan became the first geneticist ever awarded the Nobel Prize in Physiology or Medicine. As an acknowledgement of the group nature of his discovery he gave his prize money to his children along with those of Bridges and Sturtevant. Morgan declined to attend the awards ceremony in 1933, instead attending in 1934. In that time Bridges' studies of polytene chromosomes in *Drosophila* provided crucial support to Morgan's theory by confirming that genes were indeed located in chromosomes. Several of his former students and colleagues won Nobel Prizes: Hermann Muller (1946), George Beadle (1958), and Edward Lewis (1995).

Morgan's contribution to genetics is immense and his ideas and experiments transformed biology. He was an elected member of a number of American and several foreign academies. Morgan died in 1945 in Pasadena.

Source: http://en.wikipedia.org/wiki/Thomas_Hunt_Morgan

in the spatial separation of genes, offered the possibility of determining sequences in the linear dimension of a chromosome. I went home and spent most of the night (to the neglect of my undergraduate homework) in producing the first chromosome map" (Griffiths et al. 2005).

This was a great discovery. Using the same approach geneticists have been building genetic maps for nearly a hundred years. The dramatic progress in genetics would hardly be possible without genetic mapping and such grand projects as the Human Genome Project were difficult to complete. Sturtevant defined one genetic map unit (centimorgan or cM) as 1% of recombination. A similar method was reported in 1915 for building the first mammalian genetic map by John Haldane and his colleague. "This paper was written while Haldane was on active military duty in France and after his co-author, Sprunt, had been killed" (Moran and James 2005).

Thus, knowing the frequency of recombination one could build a map that shows an order and distances between genes on a chromosome. This whole approach presumes that crossing-over events are randomly

Figure 3.5 Alfred Henry Sturtevant (1891–1970), outstanding American geneticist who made the decisive contribution in the development of gene mapping and other fields of genetics. He was born in Jacksonville, Illinois, and died in Pasadena, California. (Courtesy of the Archives, California Institute of Technology.)

BOX 3.2 ALFRED HENRY STURTEVANT

Alfred Henry Sturtevant was born in Jacksonville, Illinois, in 1891. He was enrolled at Columbia University in 1908 and soon became interested in Mendelism as this could explain the traits expressed in horse pedigrees, which had been his passion since childhood. In a little while Sturtevant became an undergraduate student of Thomas Morgan and this choice determined his whole career. In 1911 following a request from Morgan, Sturtevant constructed the first genetic map. It was a great achievement that significantly affected the future of genetics.

In 1914, Sturtevant completed his doctoral work and stayed at Columbia as a research investigator for the Carnegie Institution of Washington. He formally joined Morgan's famous research team, which was exceptionally successful in experimental and theoretical genetics. Together with Morgan, Sturtevant moved to Pasadena in 1928 to work at the California Institute of Technology. He became a professor of genetics and remained at Caltech for the rest of his career.

Sturtevant work on the so-called unequal crossing-over that was important for genetics and evolutionary studies. Another bright idea was to spread gene mapping principles to "fate maps" of the fly embryo. In the 1960s this approach was resurrected, modified, and successfully used in many laboratories over the world. In honor of Sturtevant the unit of the fate maps was dubbed the "sturt" by analogy with the "centimorgan" unit of recombination mapping, which he introduced in honor of Morgan. By the early 1930s Sturtevant became the leader of a new genetics research group at Caltech, whose members included George W. Beadle, Theodosius Dobzhansky, Sterling Emerson, and Jack Schultz. The outstanding contribution to genetics made by Alfred Sturtevant put him in the category of classics. He died in Pasadena in 1970.

Source: http://en.wikipedia.org/wiki/Alfred_Sturtevant

distributed along the chromosome. If the recombination events were not random and concentrated in relatively few positions on a chromosome this method would not work. A huge amount of data accumulated so far clearly demonstrates that genetic mapping works very well indeed. Within a few years after Sturtevant's discovery it became clear that rather short genetic distances can be well approximated by this simple method, which takes into consideration only the observed frequency of recombination. Double and more complex recombination events that occur between distant genes, some of which escape direct observation, significantly affect estimates of genetic distances. To resolve this problem John Haldane (1919) suggested the so-called mapping function based on the Poisson distribution that describes the probability of random events. This mapping function is widely used and provides fairly accurate results that testify in favor of the randomness of recombination events.

Molecular nature of recombination

On the molecular level meiotic recombination is a complicated process served by many enzymes and regulated by numerous genes. The first stage of the process is the formation of the so-called early recombination nodules associated with emerging sinaptonemal complexes, special structures developing in early meiosis for the alignment, pairing, and recombination of homologous chromosomes. According to the current views double-strand breaks are initiated in some of these nodes. The early nodules are spread relatively evenly (Moens et al. 2007) and the distribution of distances between adjacent early recombination nodules is random (Anderson, Hooker, and Stack 2001). The total number of early nodules per nucleus in the mouse is 250 to 300 and declines steadily until the recombination is accomplished. The average number of meiotic recombination events per nucleus in the mouse is about thirty. It means that only a fraction of the early recombination nodules eventually becomes the point of meiotic recombination. Interference between recombination events on the same chromosome is a well-known phenomenon that makes closely located recombination events very rare. A simple reaction-diffusion physical model, where "randomly walking" precursors are immobilized and transformed into recombination points, provides a good description of this complex genetic process (Fujitani et al. 2002).

Distribution of cross-overs along chromosomes

Meiotic recombination is a double-edged sword. It can produce useful novelties but it also can destroy existing and well-adapted combinations of alleles. Optimization of such a complex and dynamic system is essential during the evolutionary process particularly when conditions change quite rapidly. There is a considerable individual variation in rate of recombination, and artificial selection is able to affect it (Brooks and Marks 1986). Differences in recombination rates can be found at the species level. The average number of chiasmata per karyotype in mammalian species studied differs at least fourfold (Dumont and Payseur 2008) and correlates well with the number of chromosome arms (Pardo-Manuel de Villena 2005). Not less than one chiasma is usually formed at each chromosome arm, that is why species with a larger number of chromosome arms have more chiasmata and vice versa. The physical length of the genome is of secondary importance for the recombination rate. In marsupials frequency of recombination usually is significantly lower than in humans and many other placental mammals despite a similar genome length. The number of chromosomes in marsupials is also smaller than in humans and this should be also a contributing factor.

The actual distribution of cross-overs along chromosomes within a species varies significantly (Froenicke et al. 2002; Borodin et al. 2008). Typically the centromeric regions of chromosomes have a lower frequency of recombination; on the contrary telomeric regions usually harbor more recombination events. The distribution of cross-overs along a chromosome is fluid and can be affected by sex, genotype, and other factors. Numerous recombination hotspots are quite typical. In the human genome, for instance, recombination spots occur on average every 200,000 base pairs preferentially outside genes (McVean et al. 2004). Assuming that a human chromosome can be as long as 200,000,000 base pairs, one can conclude that hundreds of recombination hot spots are more or less evenly distributed along a chromosome. The hot spots are not only abundant but flexible in the human genome (Jeffreys et al. 2005). Indeed, significant differences in cross-over distribution along the sinaptonemal complexes among different human individuals were documented for several chromosomes (Sun et al. 2006). Based on statistical analyses of large data sets Myers et al. (2005) proposed a two-stage model for recombination in which hot spots are stochastic features. Recombination rates within narrow regions are highly dynamic and vary among human populations, while general recombination patterns are well conserved across human populations (Serre, Nadon, and Hudson 2005).

Meiotic recombination generates randomness

Despite factors shaping the distribution of recombination events along chromosomes, there are ample opportunities for randomness. For any interval between two genes located on the same chromosome there is a probability of meiotic recombination. This probability could be high or low depending on distance between the genes, molecular features of the region, and other circumstances. However, it is completely unknowable whether a cross-over will occur in a particular region in the next meiotic cell, where exactly it will occur, and why this would happen. It is hard to know with certainty whether such knowledge cannot be obtained in principle due to randomness of the process or extreme complexity of the meiotic system, which operates using deterministic rules. Regardless of the answer to this dilemma, the conclusion can be drawn that meiotic recombination creates an additional layer of randomness in biological processes.

In the simplest case when only two genes, each with two alleles, are considered, four (2^2) potential haplotypes (combination of alleles) can be found in a population (Figure 3.6). If recombination between these two genes did not occur or some haplotypes were lost, the number of haplotypes should be smaller. In the case of three genes each with two alleles the expected number of haplotypes is eight (2^3). Different chromosomes contain different numbers of genes, from as many as two to three thousand genes per chromosome to just a few hundred genes. For a hundred genes each with two alleles the number of potential haplotypes became huge: 2^{100}. Obviously the number of individuals in a population is much smaller and the vast majority of these haplotypes do not exist. Still there

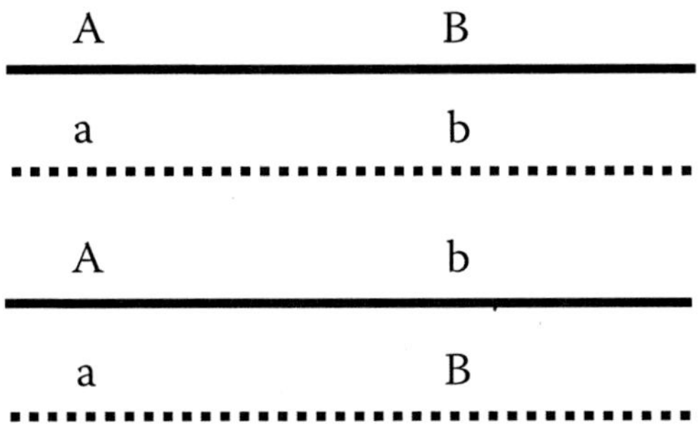

Figure 3.6 Four possible haplotypes (AB, ab, Ab, aB) that can be produced by recombination between genes A and B located on the same chromosome, assuming that each gene has two alleles.

are thousands, if not millions, of haplotypes that are constantly produced by ongoing meiotic recombination. Some haplotypes are rather common, others very rare and this fact reflects their complex dynamics affected by selection and random forces operating in the populations.

The most frequent haplotypes carrying advantageous combinations of alleles have high fitness in the current conditions. Constant reshuffling of allelic combinations is an additional source of genetic variability, which utilizes a variety of mutations accumulated in populations. Haplotypes, as Verhoeven and Simonsen (2005) put it, "develop stochastically under random recombination." This makes meiotic recombination a powerful generator of randomness that affects the evolution of species and contributes significantly to the uncertainty of life.

Origin of meiosis and sex

What is known about the origin of meiosis? Despite solid attempts to resolve this problem the answer is still elusive (Maynard Smith and Szathmáry 1995). There are, however, several groups of well-established facts which make reconstruction possible. Sex, being the quintessential eukaryotic process, most likely emerged in early unicellular eukaryotes, which prior to that were haploid, that is, had only one set of homologous chromosomes. The sexual process in eukaryotes includes production of usually haploid sperm and egg cells and their coupling, which restores diploidy (two parental copies of chromosomes in a nucleus).

While it is a common view that sex is directly related to reproduction, which is true for the majority of modern eukaryotes, it was not always the case. Some groups of unicellular eukaryotes do not have and likely never had sexual process and their reproduction is asexual (Raikov 1982). Other unicellular eukaryotes have sexual process, but their reproduction is usually asexual. Only in multicellular eukaryotes is reproduction tightly linked with the sexual process and sexual reproduction became an advantageous trait in the majority of species. The origin of the sexual process had profound biological consequences. Formation of diploidy and the inevitable haplo-diplo cycle is among the first such consequences. Without meiotic division a chromosome number should double after each successful fertilization, which is unsustainable. Segregation of homologous chromosomes during the first meiotic division has an exceptional evolutionary importance as a very powerful device randomly generating effectively an unlimited number of genetic combinations. This matter is considered in Chapter 5.

It seems plausible that the emerging sexual process instigated development of meiosis, which likely evolved from a simple one-step process to classical meiosis with complex prophase, crossing-over, and two consecutive divisions (Ruvinsky 1997). Crossing-over can be seen as a by-product of emerging meiosis. A complementary line of reasoning is related to ever occurring recombination of DNA molecules, which could be used "as an assortment process via the specific creation of double stranded breaks" for removing new mutations unavoidable during DNA duplication (Gessler and Xu 2000). This provides a view of meiosis from the perspective of DNA repair.

Regardless of the original scenario the crossing-over became an exceptionally important catalyst for eukaryotic progression due to its ability to generate random combinatorial novelties on a great scale. Another likely consequence of the developing sexual process and meiosis was transition to multicellularity, which probably took place more than once during evolution. The existence of three kingdoms of eukaryotes, including animals, fungi, and plants, with different types of sexual cycle and forms of meiosis, testifies in favor of this idea (Ruvinsky 1997).

Recombination and chromosome rearrangements

Meiotic recombination nearly always rejoins homologous chromatids. From time to time, however, rare exchanges between nonhomologous chromosomes happen. Both homologous and nonhomologous recombination may produce rearrangements. There are several types of rearrangements, including duplication of chromosome sections, inversions (turning a section of chromosome for 180°), translocations (rejoining sections of nonhomologous chromosomes), fissions (splitting one chromosome into two), and fusions (joining two nonhomologous chromosomes). Any such rearrangement is just a first step on a lengthy road toward making this novelty a permanent feature in evolving genomes. A comparison of two arbitrary species from the same taxonomic group typically reveals chromosome rearrangements, thus indicating that speciation is quite often connected to genome restructuring.

Chromosome rearrangements are usually considered as random events occurring in different chromosome points (Nadeau and Taylor 1984). More recently facts began to emerge that at least in mammalian evolution the "same regions of the genome are being broken over and over again in the course of evolution" (Alekseyev and Pevzner 2007). These facts imply the existence of so-called rearrangement hotspots. Nearly 20% of chromosome breakpoints among compared mammalian species were reused, which suggests "a high frequency of independent rearrangements occurring at the same regions of genome in different mammalian lineages" (Murphy et al. 2005). An average estimated length of the chromosome regions harboring such rearrangement hot spots is about one million base pairs. Despite relative shortness of the hot spot regions in comparison with the length of a mammalian chromosome, they are large enough to accommodate numerous independent breakpoints.

The estimates indicate that for the last ~65 million years mammalian genomes evolved at a rate of ~0.11 to 0.43 chromosome breaks per million years (Murphy et al. 2005). While these events caused a considerable transformation of mammalian genomes such chromosome rearrangements are rare. The total number of breakpoints identified so far in the mammalian genome is 492. Although the positions of breakpoints are somewhat restricted, the time of rearrangements as well as the choice of particular chromosomes involved in rearrangements seems to be mainly unlimited. The pattern of chromosome rearrangements can be described as rare, independent, and unique. This provides abundant opportunities for randomness in the sense of the book.

Genome transformations and speciation

A single recombination event producing a chromosome rearrangement in an individual is no more than a rare aberration. The majority of new rearrangements are going to be lost sooner rather than later. One of the main reasons behind this is a reduced fitness of heterozygotes for chromosome rearrangements (White 1978). Nonetheless numerous rearrangements were fixed in genomes during evolution and in some phylogenetic lineages long series of similar rearrangements were accumulated in a relatively short time. The problem was recently revisited and a set of mechanisms increasing chances of fixation of chromosome rearrangements was considered (Pardo-Manuel de Villena 2005). The relevant details are briefly discussed in Chapters 5.

Chromosomes in some mammalian species are particularly prone to fusion and provide good examples of connections between the genome rearrangements and speciation. Two instances of this kind are particularly striking. A standard genome of the house mouse, *Mus musculus domesticus*, comprised from twenty pairs of homologous

($2n = 40$) chromosomes. These chromosomes are acrocentric, that is, their centromeres are located at the very ends of the chromosomes. Some isolated mice populations in the Alps have a drastically altered structure of genomes. Such chromosome races may have as few as eleven chromosome pairs ($2n = 22$). This transformation was achieved through fusions of nonhomologous chromosomes, which are also known as Robertsonian translocations. According to the current estimates these chromosomal races diverged fairly recently during the last 5,000 to 10,000 years. It is likely that the chromosomal races have an independent origin and fixation of chromosomal rearrangements happened in rather small populations (Britton-Davidian et al. 1989). A combination of random events in small isolated populations and possible selective advantages were the driving forces causing such independent, multiple, and rapid genome transformations (Capanna and Castiglia 2004). Although hybridization between different chromosome races is possible in the laboratory, hybrids have lower fertility and this is an important initial condition for future speciation events.

Deer from the genus *Muntiacus* inhabiting South China and the neighboring states of South East Asia provide another example of dramatic and unique genome transformations mainly caused by fusions of nonhomologous chromosomes. The Indian muntjac possesses the lowest number of chromosomes in mammals; females have only six and males seven chromosomes. The Chinese muntjac on the contrary has forty-six chromosomes, which is a quite typical mammalian number. How could such drastic changes arise in a short evolutionary span of time? The muntjac chromosomes probably have "sticky ends" which elevate their fusion ability as well as likelihood of numerous chromosome rearrangements. While the tandem chromosome fusions were the predominant feature in evolution of the Indian muntjac, other chromosome rearrangements were also involved. It was suggested that genome rearrangements in the genus *Muntiacus* had started from the ancestral deer karyotype, which probably contained as many as seventy chromosomes (Yang et al. 1997). A few more rare species of Muntjac deer discovered in the past several years also have an unusually low number of chromosomes (14/13 and 8/9) and demonstrate the same chromosome reduction trend.

Genetic distances between muntjac species studied, calculated using DNA sequence differences, are quite small and the rate of speciation is slow. Chromosome isolation caused by the genome rearrangements promoted speciation in this genus. The reduction in chromosome numbers in this genus, however, was not a straightforward and uniform process and as a result a series of independent and unrelated speciation events took place (Wang and Lan 2000). Despite the leading role of the rearrangements, ecological and geographic factors contributed to the successful speciation events. There are many other eukaryotic taxons where genome reorganization and speciation were connected.

Are genome rearrangements adaptive? Both the muntjac deer and the house mouse examples do not show a clear congruence between chromosome changes and adaptive features of evolving species. Despite dramatic changes in karyotypes, morphology and physiology of the species are not apparently different. This certainly does not mean that chromosome rearrangements escape natural selection.

Heterozygotes for chromosome aberrations are under strong selective pressure as significantly lower fitness is very common. Fixations of such rearrangements in evolving genomes occur despite the negative selection. It has been known for a long time that chromosomal rearrangements serve, at least in some cases, as an important first step in speciation because of their ability to create sufficiently strong reproductive isolation (White 1978). The strength of this isolation depends on fitness of heterozygotes. The lower the fitness the stronger the reproductive isolation, which is essential for successful speciation. On the other hand low fitness of heterozygotes for chromosome rearrangements is a strong obstacle on the way to their fixation. This contradiction can be resolved by random genetic drift (see Chapter 6 for further details) capable of fixing such rearrangements regardless of their selective value. There is a great deal of randomness in the initial origin of chromosome aberrations as well as in the mechanisms of their fixation.

Intron-exon structure of eukaryotic genes: randomness again

Now let us shift attention from chromosomes to genes. The majority of eukaryotic genes are built from exons and introns. Exons are protein coding sections and introns are intermediate sections which usually do not code for proteins (Figure 3.7). After transcription of DNA from a particular gene, a lengthy mRNA molecule is produced which typically undergoes splicing. This includes cutting out introns and joining exons. The resulting mRNA molecule has an uninterrupted coding message, which can be translated to a protein after a few additional modifications. Even though a lengthy debate concerning the origin of introns has not been finalized, researchers believe that the majority of introns were inserted in genes during eukaryotic evolution and the process might have slowed significantly since the early days.

The distribution of introns in genes has been among the questions of interest for the last two decades. Introns in eukaryotic genes are located either between codons (phase 0) or within codons (phases 1 and 2). Codons are three neighboring nucleotides, which code for an amino acid.

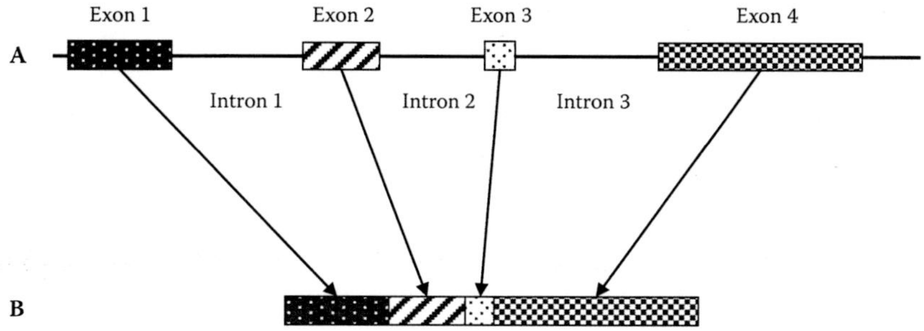

Figure 3.7 Exon-intron structure of genes and mRNA splicing. (A) The exon-intron structure of pre-splicing mRNA which mirrors the structure of a gene. It contains four protein coding exons and three noncoding introns. (B) During splicing introns cut off and exons are joined together. Spliced mRNA can be used for protein synthesis after some modifications.

Phase 0 introns are more frequent in all eukaryotes studied and phase 2 introns are less common. Several factors might contribute to this phenomenon, and frequencies of codons are among them. Some codons occur in exons much more often than others and each species has a specific set of codon frequencies. Computer simulations of the intron insertion process that generates random exon-intron sequences were carried out taking into account codon usage frequencies for a variety of eukaryotes (Ruvinsky et al. 2005). Notably in all randomly simulated data sets intron phase distribution was similar to that observed in real species with a clear bias in favor of phase 0. The complexity of the process precludes the unambiguous statement that intron insertions were random. However, it is possible to say that the simplest random description of the process is sufficient. Therefore, the core of the intron insertion process was most likely random.

Arranged randomness and immune response

Life for complex animals is impossible without a sophisticated immune system capable of defending against countless viral, bacterial, and other types of infections. Production of protein molecules, called antibodies, which are highly specific to antigens, expressed on the intruding molecules, is the core of the immune response in mammals and other vertebrates. An antibody matches to a certain antigen as a key to a lock. There is almost an unlimited number of real and potential antigens but a healthy individual who never before had any contact with an antigen is able to begin production of new and highly specific antibodies in a matter of days. How is this possible? Surely human or other genomes cannot carry millions of genes corresponding to the incredible quantity of potential antigens. Susumu Tonegawa (Nobel Prize winner, 1987) discovered the answer to this challenging question. Below is the summary of his paper published in *Nature* in 1983.

> In the genome of a germ-line cell, the genetic information for an immunoglobulin polypeptide chain is contained in multiple gene segments scattered along a chromosome. During the development of bone marrow-derived lymphocytes, these gene segments are assembled by recombination which leads to formation of complete gene. In addition, mutations are somatically introduced at a high rate into the amino-terminal region. Both somatic recombination and mutation contribute greatly to an increase in the diversity of antibody synthesized by a single organism. (Tonegawa 1983)

This summary describes the essence of the process very clearly and here we just reiterate and expand the explanation. All embryonic cells prior to differentiation of B lymphocytes, which are responsible for antibody production, have immature genes for immunoglobulins. Human chromosome 14 has a vast region which contains 86 V (variable) segments, 30 D (diverse), 9 J (joining), and 11 C (constant) segments, which are components of the immunoglobulin gene coding for the so-called heavy chain of the antibody molecule. The segments from the same category have similar length and structure but they are not identical. The variation of segments is not equal in different categories; C segments are less variable than others.

During lymphocyte development the immunoglobulin gene undergoes rearrangements until the mature version of the gene carrying only one copy of each segment emerges (Figure 3.8). These rearrangements are random and as a result each lymphocyte carries a unique gene with a new combination of V, D, J, and C segments. Effectively a section of chromosome 14 in mature lymphocytes carries DNA that is different from all other cells of an individual in which this section of chromosome 14 remains intact. This is the major deviation from standard cellular behavior during development. Another important and unique feature of the process is a very high rate of spontaneous mutations occurring in specific locations within the gene. The obvious consequence of these rearrangements and mutations is a huge variety of immunoglobulins produced by different lymphocytes.

A very similar process is happening in immunoglobulin genes coding for the so-called light chain of immunoglobulins. There are two such genes in the human genome, on chromosomes 2 and 22. Two heavy and two light chains that are joined together create a functional copy of immunoglobulin. The variety of immunoglobulins that can be produced by one individual is estimated as at least 10^8. This gigantic repertoire of

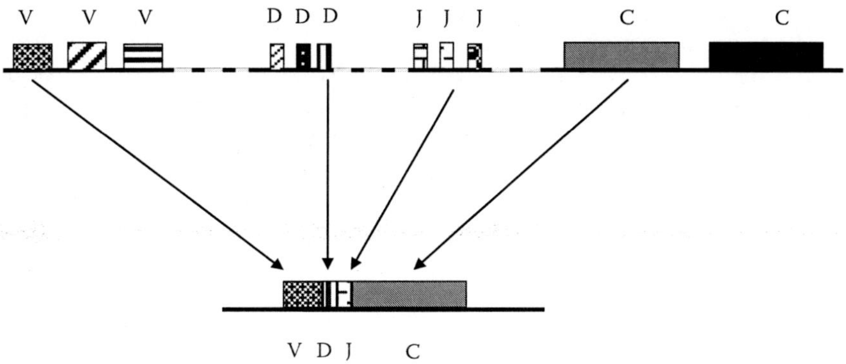

Figure 3.8 Organization of the human gene determining the heavy chain of immunoglobulin. Top: Prior to differentiation of lymphocytes the gene contains 86 V segments, 30 D, 9 J, and 11 C segments; only some of these segments are shown on the figure. Bottom: During lymphocyte differentiation a mature version of a gene containing only one V, D, J, and C segment is randomly assembled in each cell. Every lymphocyte has a unique combination of segments and produces a unique antibody.

antibodies that can be produced by a single individual using just three genes is exceptionally important for effective immune defence.

Immunoglobulins are not the only group of genes capable of generating such a massive diversity of proteins; genes coding for T-cell receptors is another example coming from the vertebrate immune system. Random recombination of components generated in both immunoglobulins and T-cell receptors is a highly adaptive feature directly related to survival. Unpredictability of forthcoming conditions is met in both cases by nearly unlimited random variability which is unravelled in each lymphocyte and in every individual. The main "objective" of this wonderful device is production of randomness in astronomic proportions. This randomness is the key strategy for adaptation and survival in an unpredictable world.

Summary

Recombination of DNA is a very ancient process. While the molecular essence of the process remains the same, namely the exchange of DNA fragments, the role of recombination dramatically evolved and diversified particularly in advanced eukaryotes. Recombination became a basis for the regular meiotic crossing-over, which significantly increased combinatorial possibilities and accelerated the rate of evolution in eukaryotes. Despite the existence of recombination hot spots and biased distribution of crossing-over events along chromosomes, crossing-over is effectively a random process. The probabilistic approach seems to be the only way to describe and predict the outcomes of meiotic recombination.

Irregular and, in a certain sense, illegitimate recombination is also widely used in genomic rearrangements. This type of recombination affects numbers and structures of chromosomes and causes reproductive isolation, the important condition of the speciation process. This type of recombination has a great degree of randomness, despite the unequal likelihood of different chromosome rearrangements.

Recombination also plays an exceptionally significant role in generating nearly unlimited variation in immune related genes. This highly adaptive feature of vertebrates evolved with the imperative to generate countless random combinations of repeated segments of the immune gene, thus ensuring quick and adequate response to any potential infection.

The ability of recombination to generate abundant random combinations of gene segments, genes, and chromosomes was widely used in evolution as the essential tool for tackling different challenges of adaptation and speciation.

References

Alekseyev, M.A., and P.A. Pevzner. 2007. Are there rearrangement hotspots in the human genome? *PLoS Comput Biol* 3:e209.

Anderson, L.K., K.D. Hooker, and S.M. Stack. 2001. The distribution of early recombination nodules on zygotene bivalents from plants. *Genetics* 159:1259–1269.

Borodin, P.M., T.V. Karamysheva, N.M. Belonogova, A.A. Torgasheva, N.B. Rubtsov, and J.B. Searle. 2008. Recombination map of the common shrew, *Sorex araneus* (Eulipotyphla, Mammalia). *Genetics* 178:621–632.

Britton-Davidian, J., J.H. Nadeau, H. Croset, and L. Thaler. 1989. Genic differentiation and origin of Robertsonian populations of the house mouse (*Mus musculus domesticus* Rutty). *Genetics Research* 53:29–44.

Brooks, L.D., and R.W. Marks. 1986. The organization of genetic variation for recombination in *Drosophila melanogaster. Genetics* 114:525–547.

Capanna, E., and R. Castiglia. 2004. Chromosomes and speciation in *Mus musculus domesticus. Cytogenetics and Genome Research* 105:375–384.

Dumont, B.L., and B.A. Payseur. 2008. Evolution of the genomic rate of recombination in mammals. *Evolution* 62:276–294.

Froenicke, L., L.K. Anderson, J. Wienberg, and T. Ashley. 2002. Male mouse recombination maps for each autosome identified by chromosome painting. *American Journal of Human Genetics* 71:1353–1368.

Fujitani, Y., S. Mori, and I. Kobayashi. 2002. A reaction-diffusion model for interference in meiotic crossing over. *Genetics* 161:365–372.

Gessler, D.D., and S. Xu. 2000. Meiosis and the evolution of recombination at low mutation rates. *Genetics* 156:449–456.

Griffiths, A.J.F., S.R. Wessler, R.C. Lewontin, W.M. Gelbart, D.T. Suzuki, and J.H. Miller. 2005. *Introduction to genetic analysis,* 8th ed. New York: W.H. Freeman.

Haldane, J.B.S. 1919. The mapping function. *Journal of Genetics* 8:299–309.

Jeffreys, A.J., R. Neumann, M. Panayi, S. Myers, and P. Donnelly. 2005. Human recombination hot spots hidden in regions of strong marker association. *Nature Genetics* 37:601–606.

Maynard Smith, J., and E. Szathmáry. 1995. *The major transitions in evolution*. New York: W.H. Freeman/Spectrum.

McVean, G.A.T., S.R. Myers, S. Hunt, P. Deloukas, D.R. Bentley, and P. Donnelly. 2004. The fine-scale structure of recombination rate variation in the human genome. *Science* 304:581–584.

Moens, P.B., E. Marcon, J.S. Shore, N. Kochakpour, and B. Spyropoulos. 2007. Initiation and regulation of interhomolog connections: Crossover and non-crossover sites along mouse synaptonemal complexes. *Journal of Cellular Science* 120:1017–1027.

Moran, C., and J.W. James. 2005. Linkage mapping. In *Mammalian genomics*, ed. A. Ruvinsky and J.A. Marshall Graves, chap. 1. Wallingford, UK: CABI Publishing.

Murphy, W.J., D.M. Larkin, A. Everts-van der Wind, G. Bourque, G. Tesler, L. Auvil, J.E. Beever, B.P. Chowdhary, F. Galibert, L. Gatzke, C. Hitte, S.N. Meyers, D. Milan, E.A. Ostrander, G. Pape, H.G. Parker, T. Raudsepp, M.B. Rogatcheva, L.B. Schook, L.C. Skow, M. Welge, J.E. Womack, S.J. O'Brien, P.A. Pevzner, and H.A. Lewin. 2005. Dynamics of mammalian chromosome evolution inferred from multispecies comparative maps. *Science* 309:613–617.

Myers, S., L. Bottolo, C. Freeman, G. McVean, and P. Donnelly. 2005. A fine-scale map of recombination rates and hotspots across the human genome. *Science* 310:321–324.

Nadeau, J.H., and B.A. Taylor. 1984. Lengths of chromosomal segments conserved since divergence of man and mouse. *Proceedings of the National Academy of Sciences USA* 81:814–818.

Pardo-Manuel de Villena, F. 2005. Evolution of the mammalian karyotype. In *Mammalian genomics*, ed. A. Ruvinsky and J.A. Marshall Graves, chap. 13. Wallingford, UK: CABI Publishing.

Raikov, I.B. 1982. *The protozoan nucleus, morphology and evolution*. Vienna: Springer Verlag.

Ruvinsky, A. 1997. Sex, meiosis and multicellularity. *Acta Biotheoretica* 45:127–141.

Ruvinsky, A., S.T. Eskesen, F.N. Eskesen, and L.D. Hurst. 2005. Can codon usage bias explain intron phase distributions and exon symmetry? *J. Mol. Evol.* 60:99–104.

Serre, D., R. Nadon, and T.J. Hudson. 2005. Large-scale recombination rate patterns are conserved among human populations. *Genome Research* 15:1547–1552.

Sun, F., M. Oliver-Bonet, T. Liehr, H. Starke, P. Turek, E. Ko, A. Rademaker, and R.H. Martin. 2006. Variation in MLH1 distribution in recombination maps for individual chromosomes from human males. *Hum. Mol. Genet.* 15:2376–2391.

Tonegawa, S. 1983. Somatic generation of antibody diversity. *Nature* 302:575–581.

Verhoeven, K.J., and K.L. Simonsen. 2005. Genomic haplotype blocks may not accurately reflect spatial variation in historic recombination intensity. *Mol. Biol. Evol.* 22:735–740.

Wang, W., and H. Lan. 2000. Rapid and parallel chromosomal number reductions in muntjac deer inferred from mitochondrial DNA phylogeny. *Mol. Biol. Evol.* 17:1326–1333.

White, M.J.D. 1978. Chromosomal modes of speciation. In *Modes of speciation*. San Francisco: W.H. Freeman.

Yang, F., P.C. O'Brien, J. Wienberg, and M.A. Ferguson-Smith. 1997. A reappraisal of the tandem fusion theory of karyotype evolution in Indian muntjac using chromosome painting. *Chromosome Research* 5:109–117.

chapter four

Uncertainty of development

> Whether germ cells succeed in making eggs or sperm depends both on their genetic constitution and on the tissue environment in which they develop.
>
> **Ann McLaren**
> *Germ Cells and Germ Cell Sex*
> (1995)

Phenotype and genotype

At the dawn of genetics in 1865 Gregor Mendel made two fundamental discoveries. The first was that each gene is represented by two copies in an organism, and this conclusion was proven to be true for the vast majority of eukaryotic species. The second drew a distinction between the genetic constitution of an individual and the expressed traits. This conclusion was also supported by subsequent progress in genetics. In order to properly describe these observations the special terms *gene* and *allele,* as well as *genotype* and *phenotype,* were introduced by the Danish scientist Wilhelm Johannsen as early as 1905. William Batson, leading British scientist and a towering figure in the early twentieth century, independently introduced the terms *gene* and *genetics* at the same time.

The need for the terms *genotype* and *phenotype* stems from the simple and well-known facts that individuals with the same phenotypes may have different genotypes (*AA* and *Aa*) and, vice versa, individuals with the same genotype may have quite different phenotypes. The former case is usually explained by the presence of a dominant allele in heterozygotes and the latter by variable expressivity or low penetrance of different genotypes. These two terms were finally introduced into the genetic lexicon by Nikolay Timofeeff-Ressovsky in 1926 (Timofeeff-Ressovsky and Timofeeff-Ressovsky 1926).

Depending on the situation the term *phenotype* refers to one or many traits of an organism. It is hardly possible, however, to provide a complete description of an individual phenotype because that individual may possess myriads of phenotypes and many of them escape observation, identification, and measurement. In addition an individual's phenotype is constantly changing from the initial zygote to the final stages of ontogenesis. Thus, the term phenotype embraces all sorts of traits, including those that characterize molecular and cellular organization, morphology, physiology, behavior, etc.

Genotype seems to be a more concrete term. It usually describes the genetic constitution of an individual, with specific reference to a characteristic under consideration. For a simple qualitative trait, genotype defines a certain allelic combination that can determine a phenotype. For instance, *aa* would mean black mouse, while *Aa* would mean grey mouse. However, for many traits, such simple description is not sufficient. There are several or many independent factors affecting the development of phenotypes and thus increasing distance between genotype and phenotype. For numerous quantitative traits, like weight or height, all these considerations are particularly relevant. This is why saying that the genotype of a mouse or fly is *Aa* may not be sufficient to predict phenotype. For a more realistic description one should know either the degree of expression of a trait (expressivity), for instance, longer or shorter tail, or proportion of individuals (penetrance) with the same genotype that will carry a mutant or wild trait, for instance, straight or curled wings.

Thus, mutant mice having the same formal genotype may show tail length variation say from 90% to 27% of an average normal individual. The trouble is that normal individuals also demonstrate a significant variation of the trait. The best one can do is to measure the phenotypes of many normal and mutant individuals and provide a correct statistical description of the trait's expressivity. Now let us imagine that some individuals with the same genotype express a mutant trait while others do not. It could be a developmental syndrome or anything else. If seventy individuals from every hundred that have the same genotype manifest the trait, this means that penetrance of the trait is 70%.

It has been proven in countless experiments and observations that phenotypes are very much influenced by different factors and cannot be predicted exactly in the majority of situations. In the textbook examples phenotypes can be predicted easily, if genotypes are known (Figure 4.1). The question remains, what is the proportion of these illustrative examples? Scientific practice indicates that such instances are not very common. For a long time geneticists used numerous modifications of the following formula in order to describe relations between genotype and phenotype:

$$\text{Genotype} + \text{Environment} + \text{Random Variation} \rightarrow \text{Phenotype}$$

In fact there is a multitude of reasons that the road from genotype to phenotype is very long and convoluted. Many of these reasons are probabilistic and as a result the exact prediction of phenotype, even if the genotype is precisely known, is impossible. In this chapter we shall concentrate attention on the contribution of random factors to developmental processes. Environmental factors are omitted from these considerations

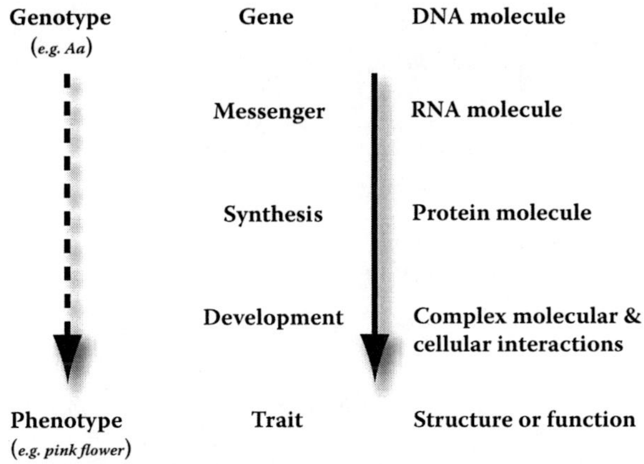

Figure 4.1 Classical view on transferring hereditary information from genotype to phenotype. There is a great deal of uncertainty on the way from a gene to a trait caused by random events and complexity of the molecular and cellular processes. Even precise knowledge of genotype is not sufficient for exact prediction of phenotype.

as it is quite clear that due to the stochastic nature their contribution to developmental processes greatly elevate randomness.

Stochasticity of development: clones and twins

Let us consider an imaginary experiment. A zygote undergoes individual development and a phenotype is measured. Then the experiment is repeated again and again with an exactly identical zygote under the same conditions. Do we expect to observe the same phenotype in each experiment? As the formula above hints, we probably do not. And if so, this argues in favor of randomness during development. Now we shall try to find suitable experimental circumstances when a similar experiment could be emulated. One has to accept that creation of exactly the same environment for numerous developments might not be practically achievable. It is also true that rare spontaneous mutations as well as recombination events will inevitably occur during development and bring some changes in initially identical genotypes. Nonetheless, situations similar to the imaginary experiment can be found and we are going to consider two of them in the following material.

Clones of parthenogenetically reproduced animals create a close imitation of the required conditions. It became nearly customary to call this type of reproduction asexual, which is certainly incorrect. It has to be remembered that in parthenogenetic populations females produce eggs that undergo development without fertilization. True asexual populations

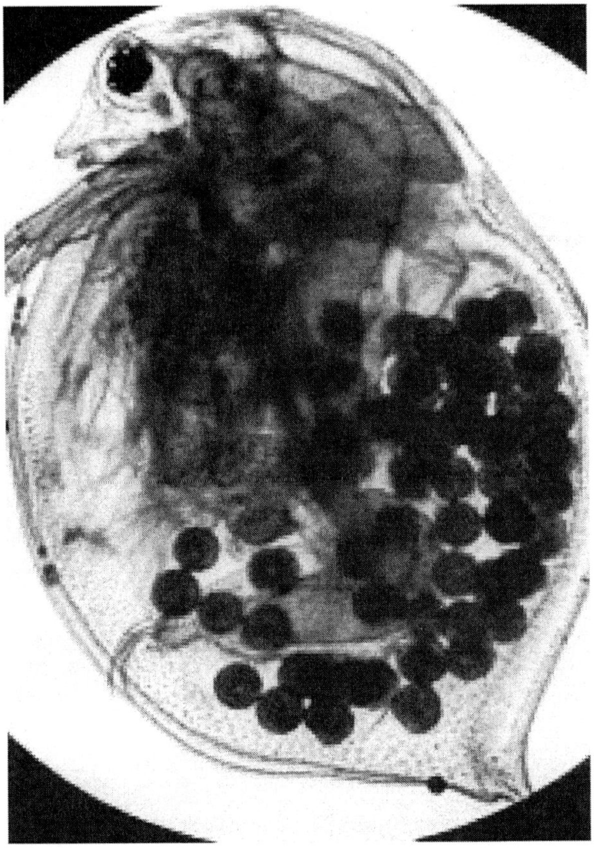

Figure 4.2 The waterflea, *Daphnia pulex*. These small freshwater crustaceans can reproduce parthenogenetically, that is, unfertilized females produce diploid eggs. Such mitotically produced eggs are nearly identical and develop without fertilization in highly uniform conditions in the brood pouch inside the carapace. A female with numerous parthenogenetically produced embryos is shown here. http://en.wikipedia.org/wiki/Daphnia. (Photo courtesy of USGS, La Crosse, Wisconsin.)

have neither males nor females. The small freshwater crustacean *Daphnia pulex*, the water flea, is a good example (Figure 4.2). The most typical mode of their reproduction is ameiotic parthenogenesis, when females produce diploid eggs by mitotic divisions and thus avoid meiosis and subsequent fertilization. Such eggs produced by a single female are expected to be nearly identical and they develop in highly homogeneous conditions within the same brood pouch. Rare spontaneous mutations may cause some deviations among offspring but they should be rather uncommon as the number of cells in an organism is not too large and the life cycle is short. Progeny from the same female create a parthenogenetic clone, a group of individuals that are nearly identical from the genetic point of view. After hatching the young water fleas belonging to the same clone can be grown in a jar with fresh water, where conditions such as temperature, light intensity, and food availability are very much similar for all individuals.

Despite the expectations, significant phenotypic variations in body size and shape can be easily observed several days after the hatching. It means that even the most similar genetic constitution and practically the same environmental conditions do not necessarily lead to uniformity of phenotypes within a clone. This observation also raises the question about factors other than genetic constitution and environment that promote phenotypic variations. Interestingly not only morphological traits, which are the result of complex interactions and are distant from the primary gene products, show variation in parthenogenetic clones. It was found that even those traits that are the closest to genes, like electrophoretic mobility of enzyme glucose 6-phosphate dehydrogenase (G6PD), varied within *Daphnia pulex* clones (Ruvinsky, Lobkov, and Belyeav 1983). These variations are unlikely to have resulted from new mutations, which do not occur so frequently during development. Despite a lack of obvious genetic variation within a clone the frequency of individuals with different G6PD electrophoretic variants can be changed by selection. This fact tentatively indicates the possibility of such nonmutational variation being transmitted from generation to generation within ameiotic parthenogenetic clones. As shown in mice, modified phenotypes in the following generation could be caused by the transfer of certain forms of mRNA through gametes to zygote (Rassoulzadegan et al. 2006). This type of epigenetic inheritance, which usually plays an important role during individual development, may in some cases spill over to the next generation.

Another example of clonal variability was recently found in parthenogenetic marbled crayfish (Vogt et al. 2008). Animals from the same clone that were shown to be isogenic by microsatellite analysis demonstrated a broad range of variation in color, growth, life span, reproduction, behavior, and a number of sense organs, even when reared under nearly identical conditions. The authors concluded that such developmental variability can introduce randomness into life histories, eventually modifying individual fitness and population dynamics.

A less exotic example of intraclonal variation can be found by studying human monozygotic twins, which originate from the same zygote and hence have identical genome. Of course in this case a clone typically consists of two individuals only. If such embryos develop in similar conditions they usually have quite similar phenotypes. However, many traits, such as fingerprints, are never the same. For years hundreds of studies revealed discordant traits in monozygotic twins. More complex traits, like susceptibility to some diseases, are particularly common among them. What are the causes of the differences? So far several groups of factors have been identified that affect unravelling genetic information during development and hence influence formation of phenotype. These factors include mosaicism, epigenetic variability, alternative splicing, stochasticity of gene activity, and other phenomena. Their individual and combined contribution affects development not only in twins or members of the same clones but in any individual. One could say that these forces "divert" development from an "ideal" pass proscribed in the genome. However, this seems to be a naive outlook because individual development always occurs in a "real world" situation and random factors are an integral part of it. In other words the influence of random factors is unavoidable and the question is which factors would affect developmental processes and how. The accumulated result of these influences, considering a pair of identical twins, is expected to be more pronounced for longer and more intensive influences. Indeed, twins often become less alike as they age, which of course in part can be explained by external influences (Martin 2005). Thus, twins and parthenogenetic clones provide very suitable models for assessing the contribution of random factors in developmental processes.

Veitia (2005) considered the question of stochasticity versus the fatal "imperfection" of cloning. No doubt rare mutations, which usually occur due to replication, recombination, or repair errors, contribute regularly to variations within a clone. However, more significant contributors are stochastic events constantly affecting development and causing phenotypic variations that are not determined by genome. The following sections of this chapter concentrate on different sources of stochastic events and randomness.

Mosaics and chimeras

A chimera is an organism that is composed of two or more genetically different cell populations that originated from distinct embryos. Natural chimeras result from the exchange of cells between developing embryos and occur rarely; their investigation is difficult. On the contrary experimentally produced chimeras provide excellent opportunities for studying numerous biological problems related to development (Le Douarian and McLaren 1984). Mosaicism is quite similar to chimerism and is also used successfully for investigating development (Yochem and Herman 2003). The major difference is that the dissimilar cell populations in mosaics emerged from the same zygote. Mosaicism occurs regularly in multicellular organisms despite the low rate of new mutations and this can be explained by the large number of cells and cellular divisions. If such mutations arise early enough during development, two or more cell lines may exist in tissues or organs of an individual and affect the phenotype. Such somatic mosaicism may lead to variable expressivity of some traits or even can affect penetrance of a trait. In these cases the expected phenotype proscribed by the genotype does not develop. Even "simple" monogenic traits often behave unexpectedly due to mosaicism (Gottlieb, Beitel, and

Trifiro 2001) or chimerism (Redina et al. 1994). The frequency distribution of cell types in chimeras is rather uniform, whereas in mosaics it is binomial (Falconer and Avery 1978). This explains greater variability in chimeras. A specific but very important case of mosaicism, random inactivation of the X chromosome in mammalian females, is considered later.

Transposable or mobile genetic elements, first discovered by Barbara McClintock in the 1940s (McClintock 1953), who won the Nobel Prize in 1983, is another source of mosaicism (Nitasaka and Yamazaki 1991). Mobile elements change genome locations during development in seemingly random fashion (Kazazian 2004) and by doing so cause mutations, chromosome rearrangements, and also affect the activity of neighboring genes. All these effects generated by mobile elements create distinct cell lines and augment the frequency of mosaicism.

Mutations and transpositions of mobile elements being random factors can only intensify the uncertainty of development by leading to mosaicism. Dozens of genetically determined syndromes and diseases exhibiting variable expressivity have been linked to somatic mosaicism (Gottlieb, Beitel, and Trifiro 2001) and support its significance as a common developmental phenomenon. Mosaicism and chimerism increase the uncertainty of phenotype formation and further remove phenotype from genotype.

Alternative splicing and variety of proteins

Now we focus attention on an entirely different form of mosaicism, observed at the mRNA level, called alternative splicing. As briefly explained in Chapter 3, the majority of eukaryotic genes consists of protein coding parts, exons, and intermediate parts called introns (Chapter 3, Figure 3.8). During RNA processing introns are excised and exons are spliced together. The process of splicing might be straightforward, when all introns are removed and all exons are joined in sequential order. However, there are numerous possibilities for deviations from this simple scenario, some of which are shown in Figure 4.3. For instance, this may include simple deviations, such as skipping some exons in the mature mRNA. More complex deviations are based on so-called mutually exclusive exons, alternative borders of exons, and other modifications. As a result more than one mRNA can be produced by the same gene; hence, more than one protein can be coded by a single gene. Some genes have exon-intron structures, which allow production of a multitude of isoforms. According to current estimates the *Dscam* gene in *Drosophila melanogaster* may encode 38,016 isoforms of mRNA and hence proteins through extensive alternative splicing. The 95 alternative exons in this gene are organized into clusters that are spliced in a mutually exclusive manner (Celotto and Graveley 2001). *Dscam* has an important role in neural wiring and pathogen recognition

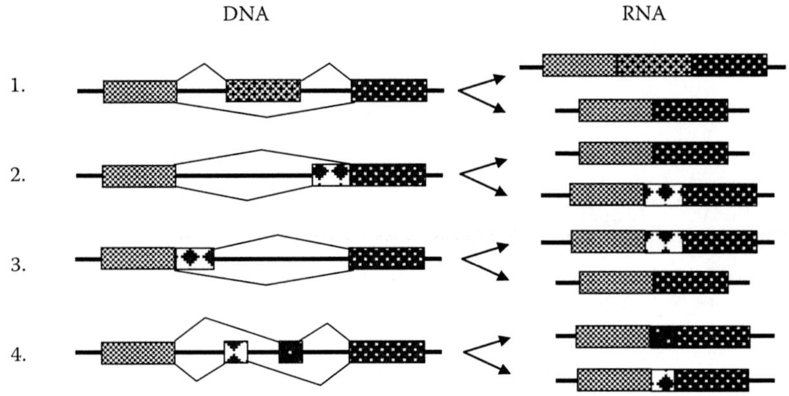

Figure 4.3 Several forms of alternative splicing. Exon-intron structures of genes are shown on the left and spliced mRNA on the right (two alternative forms are shown for each considered case). (1) Skipping the second exon. (2) Alternative 5′ end of the second exon. (3) Alternative 3′ end of the first exon. (4). Mutually exclusive second and third exons. Some genes have dozens of exons and introns and combinatorial possibilities for alternative splicing become overwhelming.

and the expected number of isoforms potentially encoded by this single gene is approximately twice larger than the total number of genes in the *Drosophila* genome (Olson et al. 2007).

According to existing estimates 35% to 60% of human genes are involved in alternative splicing. Obviously this kind of molecular mosaicism creates immense coding opportunities and plasticity. In many situations alternative splicing is a highly regulated process that leads to production of different isoforms in different cell types or tissues. The functional importance of alternative splicing is undoubtable in these cases. There is, however, a great deal of evidence that mutations in exons and probably introns lead to nonfunctional forms of alternative mRNA splicing causing significant phenotypic deviations and diseases (Cartegni, Chew, and Krainer 2002). For the majority of these isoforms there are very limited indications that they could be properly functional as mRNAs or proteins (Tress et al. 2007). Some of these alternative mRNAs and proteins compete with the regular isoforms and such interactions affect development and phenotype formation. The mutation *Fused* in mice is an example of how alternative splicing caused by insertion of a mobile element into intron 6 of the *Axin* gene changes tail phenotype from straight and long to short and kinked (Ruvinsky et al. 2000). Expressivity of this mutant allele varies significantly (Figure 4.4) and correlates positively with the quantity of alternatively spliced mRNA (Flood and Ruvinsky 2001).

On the other hand it was found that sections of genes with high incidence of alternative splicing events usually (~81%) correspond

Figure 4.4 Variable expressivity of tail abnormalities in mice heterozygous for *Axin*^{Fu} mutation: (top) strong abnormality; (bottom) nearly normal tail. There is a positive correlation between the quantity of alternatively spliced mRNA and tail abnormalities.

to intrinsically disordered regions in proteins (Romero et al. 2006). Such protein regions, as a rule, do not belong to the typical folded structures of protein molecules and have a considerable degree of flexibility. Interestingly these regions play a role in signalling and regulation and this might be an essential factor in the speedy evolution of multicellular eukaryotes. This observation demonstrates that structural and/or functional uncertainty is essential for integration of the complex molecular networks discussed below. It is likely that other regions of the eukaryotic genes are more effectively protected from alternative splicing events by natural selection, which removes mutations generating alternative splicing.

Sometimes splicing sites are shifted by three nucleotides, which is the result of stochastic binding of the molecular complexes responsible for splicing. The presence of alternative splice sites located nearby accelerates mistakes of the splicing machinery (Chern et al. 2006). While the alternative splicing has a clear functional role and is well controlled in many situations, there are numerous indications that randomness and stochasticity are constant and frequent features of this complex process. If so, alternative splicing is an additional source of developmental noise. This may bring opportunities and disadvantages to those eukaryotic species in which alternative splicing is common.

Stochastic nature of gene activity

The fundamental cellular processes like mRNA and protein synthesis are essentially very complex chemical reactions quite often with relatively low copy numbers of involved molecules. Stochastic fluctuations in these and other basic processes within cells are very much expected. Indeed, it was reported years ago that gene activations are stochastic (Ko 1992). Experimental and theoretical investigations of such fluctuations, particularly in a single cell, advanced considerably during the last decade. It became more obvious that random fluctuations in gene expression generate cellular variability. Raser and O'Shea (2004) measured the differences in expression of two alleles in a diploid cell and estimated the noise intrinsic to gene expression. This noise is gene specific and not dependent on the regulatory pathway or absolute rate of expression. Investigation of mutations affecting the noise level indicates that the noise can be optimized and, if so, this represents an evolvable trait.

Studies of transcription rate in individual cells have shown that the amplitude of noise is under the control of genetic factors (Elowitz et al. 2002). Stochasticity observed in cells has both intrinsic and extrinsic components. The first one is inherent in biochemical processes such as transcription and the second reflects fluctuations in the cellular components involved. Three categories of gene expression noise have been defined: gene-intrinsic, network-intrinsic, and cell-intrinsic (Kærn et al. 2005). The observed fluctuations in gene activity are not necessarily due only to the low number of molecules expressed from a gene but rather originate in the random events of gene activation. Transcriptional regulators, which themself are products of random expression, propagate noise to their target genes (Becskei, Kaufmann, and van Oudenaarden 2005). A stochastic model of gene expression in response to an inducing signal gives a fairly accurate description of the experimental facts. A conclusion can be drawn that the transitions between the active and inactive states of a gene are random in nature (Karmakar and Bose 2004).

The intensity of the noise or stochastic fluctuations may vary but it is a permanently existing factor. The question of biological relevance of stochastic fluctuations in gene expression was investigated and the predictions were confirmed that production of essential and complex-forming proteins is characterized by a lower level of noise than production of other proteins (Fraser et al. 2004). This finding supports the notion that noise in gene expression is a biologically important parameter and it is likely controlled by natural selection. We shall return to this question in the section devoted to gene networks and canalization. The analytical framework for investigation of the theoretical aspects of stochastic fluctuations is advancing well (Levine and Hwa 2007).

Stochasticity, which is a typical feature of simple molecular interactions, penetrates well into the basic molecular biological processes such as transcription and translation. Gene expression fluctuates constantly and does so differently in different cells. Even transcription levels of two identical alleles in the same cell might differ. The results of numerous studies unequivocally demonstrate that randomness is an intrinsic property of gene expression and most likely other intracellular processes. The intensity of this molecular noise depends on many factors and can be instantaneously modulated in individuals by external factors as well as steadily evolve in populations under control of natural selection. The external noise only exacerbates randomness of intracellular processes and, hence, in a wide range of situations quantitatively precise predictions of cellular behavior become even less attainable. In multicellular organisms stochasticity of gene activity is a major contributor to differences between cells of the same type, and this is an important contributing factor that leads to the discordance observed in monozygotic twins or individuals belonging to the same clone.

Epigenetic basis of developmental variability

Epigenetic studies have a long and convoluted past. The term *epigenetics* was introduced by C.H. Waddington in the 1940s prior to the emergence of molecular biology. It has been defined more recently as "temporal and spatial control of gene activity during the development of complex organisms" (Holiday 1990). This definition covers a wide range of research directions relevant to the regulation of gene activity, from switching genes on and off to cell differentiation and phenotype development.

Genome studies made it apparent that even precise knowledge of the primary structure of DNA is not sufficient for understanding which genes are active or inactive at a particular time and why (Whitelaw and Garrick 2005). Activation and inactivation of genes is critically important for the vast majority of developmental processes and in many situations is achieved by modification of DNA. Methylation of cytosine is a typical way of epigenomic DNA modifications that determine the status of gene activity. Transcriptional activity of highly methylated genes is usually suppressed and conversely demethylation activates genes.

Chromatin, being the essential component of eukaryotic chromosomes, is a complex structure composed of DNA, proteins, and other molecules. Changes in chromatin structure have major repercussions for gene activity. Histones, which are proteins particularly important for DNA packaging, play a key role in structuring chromatin. Chemical modifications of histones affect chromatin structure and hence gene activity. There are several chemical processes of histone modification, including acetylation, phosphorylation, and methylation (Whitelaw and Garrick 2005). Chromatin remodeling connects chemical modifications of histones and DNA methylation with gene activity, thus making these biochemical processes critically important for epigenetic regulation of gene activity and development (Verger and Crossley 2005).

Epigenetic signals, once established during development, can be transmitted to the next mitotically produced generations of somatic cells with significant fidelity. This is why cells from different tissues maintain their identity and functions. As earlier mentioned heavy methylation of certain DNA regions leads to gene silencing. However, much less is known about why the methylation signals are put in particular regions and what initiates the process. Deviations are inevitable in all biological processes, including establishment and maintenance of epigenetic marks. Not surprisingly many epigenetic signals may be missed or misplaced and this causes "epimutations" affecting phenotype. This is particularly common for situations when conditions are far from normal and epimutations are more frequent. Indeed, embryonic stem (ES) cells used for experimental cloning by nuclear transfer have demonstrated that "the epigenetic state of the ES cell genome is extremely unstable" (Humpherys et al. 2001). Certainly epigenetic perturbations caused by cloning must be very significant. This explains the inefficiency of cloning by nuclear transfer as the majority of embryos die before birth and rare survivors display growth abnormalities.

Monozygotic twins represent a good opportunity for studying epigenetic modifications. It was found that the number of epigenetic differences between identical twins increases with age. For instance, fifty-year-old twins have over three times the epigenetic difference of three-year-old twins. "These findings indicate how an appreciation of epigenetics is missing from our understanding of how different phenotypes can be originated from the same genotype" (Fraga et al. 2005). A similar notion is expressed in the recent review of the problem (Petronis 2006). It seems very likely that the lack of stability in maintaining epigenetic signals is a consequence of stochastic changes that occur during development.

In the previous section we discussed the stochasticity of gene expression, which can be regarded as a part of the epigenetic transformation of hereditary information stored in DNA molecules into phenotypes. Here we are going to pay more attention to switches between active and inactive states of genes, which is commonly considered as a typical outcome of an epigenetic process. Another objective of this section is discussion of random activation and inactivation of genes in development. Both processes contribute to uncertainty in development and increasing distance between genotypes and phenotypes.

Methylation of DNA has been considered an important mechanism of regulation of gene activity since the seminal paper of Holliday and Pugh (1975). Heavily methylated genes, as earlier mentioned, are usually silenced. It seems that this mechanism of gene regulation gained significant importance during the evolution of mammals, while in other eukaryotes its role is less pronounced. One of the ways to assess stability of methylation signals and some consequences is to study deviations from a standard pattern. Beckwith–Wiedemann syndrome (BWS) in humans is a disorder characterized by developmental abnormalities and predisposition to childhood tumors. A cluster of relevant genes located on chromosome 11 normally undergo parental imprinting, the phenomenon when only one allele of a gene in the cluster is active and the second is located in a homologous chromosome. Activation of the second allele is the cause for BWS. Many identical twin pairs, mostly female, have been reported to be discordant for BWS. It is very unlikely that mutations in twins could occur so frequently and increase discordance as significantly as it was observed. The methylation pattern of the *KCNQ1OT1* gene, which is critically important in this phenomenon, provides a good explanation for BWS (Bestor 2003; Figure 4.5). The maternal allele of the gene is normally silenced and the paternal allele is expressed. Methylation marks are usually maintained on both DNA strands of the maternal allele. If one of these marks was lost prior to twinning, one twin will get a silenced maternal allele methylated on both strands and the second twin will get an unmethylated and hence active allele. This second twin will have two active *KCNQ1OT1* alleles and develop BWS. The loss of imprinting at *KCNQT1OT1* might not always be purely stochastic (Bestor 2003), but in the majority of situations this seems to be the most likely explanation. Interestingly RNA produced by *KCNQT1OT1* does not code a protein and possibly can be used in regulation of activity of other genes. Many imprinted genes also produce protein noncoding RNA, which is probably involved in gene regulation.

Activation and inactivation of genes is typically governed by transcriptional factors, products of other genes, which have to be regulated as well. If this logic is correct, an effectively infinite number of regulators

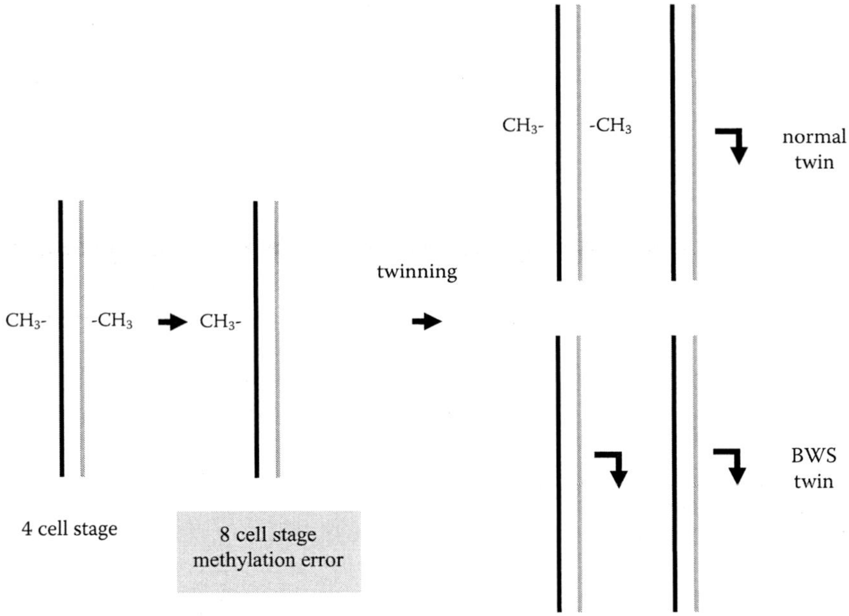

Figure 4.5 A simplified model proposed by Bestor (2003) for high incidence of Beckwith–Wiedemann syndrome (BWS) in one of two monozygotic (predominantly female) twins caused by the methylation error. In humans some genes undergo imprinting, a situation in which only one allele remains active and another allele is silenced by the DNA methylation. A deviation from this arrangement may lead to developmental abnormalities. At the four cell stage both DNA strands of maternal *KCNQ1OT1* allele are methylated (-CH_3 mark); the paternal allele is not shown at the four and eight cell stages. Prior to the eight cell stage and the following twinning event the methylation signal was lost on one strand. Eventually one twin obtains the maternal allele with both DNA strands properly methylated, while another twin receives the unmethylated allele. The twin that gets both maternal and paternal unmethylated active alleles (shown by a downward arrow) develops BWS, causing significant abnormalities. (Redrawn from Bestor 2003, with modifications.)

is required. How is this problem resolved in real life? A comprehensive answer is not known as yet, but some researchers believe that many transcription factors are regulated at a level other then transcriptional, for instance at post-transcriptional levels. It is also a possibility that transcription factors are not regulated at all during early stages of development but rather expressed stochastically. By doing so, they determine the fates of particular cell lines when their chance expression coincides with a particular external growth factor (Verger and Crossley 2005). Theoretical analysis of the problem shows that stochastic switching may have a selective advantage over the switching occurring as a response to external changes, at least in certain conditions (Kussel and Leibler 2005; Zhuravel and Kærn 2005).

Discordant traits in monozygotic twins, as already mentioned, represent an excellent opportunity for investigation of epigenetic variability. It has been known for a long time that the different environmental influences to which twins are exposed and rare spontaneous somatic mutations are not sufficient to explain the majority of the cases. New molecular technologies bring a large volume of additional information and improve the current understanding of discordance. This includes recent studies discovering considerable differences in methylation patterns of genes affecting discordant traits in monozygotic twins (Healey et al. 2001; Tsujita et al. 1998, Petronis et al. 2003, Whitelaw and Whitelaw 2006). The average efficiency of methylation maintenance of some alleles was estimated as 0.96 per site per cell division (Chen and Riggs 2005). This estimate indicates that epigenetic fidelity is dramatically lower than in basic molecular biological processes such as transcription and translation, where mistakes are very rare. If this estimate is correct for many other genes, no wonder that in complex developmental processes involving huge numbers of cells stochasticity is very high. Epigenetic variability was also found in human germ cells indicating that some epigenetic patterns can be transmitted across generations and affect phenotypes (Martin, Ward, and Suter 2005; Flanagan et al. 2006). Researchers quite often mention that epigenetic phenomena tend to be stochastic, reversible, and mosaic. DNA methylation is not the only mechanism of epigenetic changes. Chemical modifications of histones, proteins playing a key role in chromatin remodeling, also demonstrate stochasticity. It was shown that such stochasticity resulted in establishment of the stability of epigenetic systems and provides two essential outcomes: active or inactive states of gene activity (Dodd et al. 2007). A mathematically based theory of epigenes, epigenetic hereditary units having not less than two functioning regimes, was initially proposed in 1970s and significantly developed since then (Tchuraev 2006).

Cancer, mental disorders, and aging also offer numerous examples of stochastic and environmentally caused epigenetic events (Hoffman and Schulz 2005; Ting, McGarvey, and Baylin 2006; Petronis et al. 2003, Holliday 2005). Future investigations in this field as well as emerging computational epigenetics look promising (Bock and Lengauer 2008).

Random gene inactivation events

In mammals there are three known types of gene dosage control, and parental origin-specific imprinting, which we briefly discussed in the previous section, is one of them. The essence of imprinting is switching off one of two alleles in some genes depending on its parental origin. Such imprinted alleles become transcriptionally inactive in early embryonic development and maintain their inactive status during the entire life of an individual. Inactivation failure or unexpected reactivation of such alleles causes significant abnormalities or even death. The process of inactivation of a parental allele in the imprinted genes is strictly controlled; nevertheless, imprinting stochastic mistakes as we saw are quite common. There are also two other types of dosage compensation when inactivation of an allele is necessary and the process of inactivation is completely random. One of them is the famous random X chromosome inactivation in mammalian females, discussed later in this chapter, and random autosomal inactivation, described below. Several examples of this kind have been found so far.

Several genes located on mouse chromosome 17 show monoallelic expression in bone marrow stromal cells coupled with 50% methylation level (Sano et al. 2001). Alleles of these genes switch randomly between active and inactive states during the formation of daughter cells. Olfactory receptor genes provide another example of random inactivation. In a sensory neuron only one allele from a family of 1000 olfactory receptor genes is usually expressed (Chess et al. 1994). This inactivation dramatically narrows and specifies the function of a neuron for the rest of its life. Such random and very stable inactivation, however, can be cancelled by successful cloning when nuclei of olfactory sensory neurons are inserted into oocytes. These results proved that reprogramming of the genes in terminally differentiated neurons is possible, at least in the experimental conditions, and the random inactivation does not cause irreversible changes in DNA of olfactory neurons (Eggan et al. 2004).

The latest investigations revealed that monoallelic expression with random choice between the maternal and paternal alleles is more widespread than initially thought. More than 300 genes from 4000 genes studied are subject to random monoallelic expression (Gimelbrant et al. 2007). Stochastic monoallelic expression of cytokine genes is another example of a similar kind. While in many of the cases a clear biological rational for the predominant and stochastic monoalelic expression is not known as yet, such genes are not just rare exceptions from the standard biallelic expression (Paixão et al. 2007). Hundreds of genes behaving in such random fashion can undoubtedly create very considerable developmental variability and uncertainty and if so contribute to phenotypic differences of individuals with identical initial genotypes. The essence of the evolutionary process that affected expression of these genes was formation of a genetic device capable of generating randomness. It was suggested that "random (in)activation events are common in genes and gene clusters with a low probability of transcription" and this phenomenon can be used by natural selection for creation systems with more stable monoallelic expression like autosomal parental imprinting and random X chromosome inactivation (Ohlsson, Paldi, and Graves 2001).

Random X chromosome inactivation

In 1961 Mary Lyon suggested the hypothesis that one X chromosome is inactivated in each cell of placental mammalian females. This idea was firmly supported by an immense number of publications and became universally accepted. After fertilization in those zygotes that carry two X chromosomes and are destined to become females, the maternal X chromosome is active while the paternal X chromosome is inactive. At the time of the split between cells of the future embryo (the inner cell mass) and extra-embryonic tissues (trophoblast), which serve the developing embryo, the paternal X chromosome remains inactive. This explains why in trophoblast cells the paternal X chromosome is switched off and the maternal X chromosome is switched on. Soon after this the paternal chromosome is eventually activated in the inner cell mass; that is, the embryo proper, and both X chromosomes become active for a short time. Then random inactivation of one of the two X chromosomes occurs. From this moment during development all mammalian females become mosaics because sets of alleles located at the X chromosomes inherited from mother and father are different. This complex arrangement evolved after separation of placental mammals from other mammalian species such as marsupials. The central point of our interest here is random inactivation of the X chromosome, which is a highly regulated process.

Random inactivation of one X chromosome in females is a complex stepwise process which includes counting X chromosomes, choice initiation, establishment of inactivation and its maintenance (Boumil and Lee 2001). All these steps are controlled by a group of genetic elements, the so-called X inactivation center (*Xic*) compactly located on the X chromosome. The number of X chromosomes per haploid set is assessed during the counting step. If a diploid cell has two X chromosomes, one of them is randomly inactivated in somatic cells. In diploid cells with more than two X chromosomes all of them, except one, are inactivated. In a normal diploid female cell one chromosome is committed to inactivation during the choice step. Genetic elements involved in these steps have been identified. Recent findings support a stochastic model for the counting and choice steps (Monkhorst et al. 2008).

The number of cells in the embryo when X chromosome inactivation takes place is more than thirty. If the probability of inactivation for either X chromosome is equal (~0.5) and the number of cells that undergo random inactivation of the X chromosome is at least thirty, then the probability of independent inactivation of the same chromosome (paternal or maternal) in all cells is extremely low ($P < 2^{-30}$). According to the generally accepted definition the distribution of cells is considered skewed if the same X chromosome (paternal or maternal) is presented in 75% of cells. The distribution is considered as highly skewed if 90% of cells or more have the same inactive X chromosome. The process of random inactivation can be modeled by Bernoully trials with two equal outcomes ($P = 0.5$) and the number of trials equal to thirty. Assuming random inactivation individuals with skewed distribution are expected in less than 1% of females and those that have highly skewed distribution in less than 0.0004%. The frequency of females that have skewered distribution is considerably higher even at birth and increases further with age at least in some tissues (Bolduc et al. 2008). This finding raises two questions: why the observed distribution is more skewed than expected and why aging affects skewness.

There are two explanations of this fact (Figure 4.6, Minks, Robinson, and Brown 2008). First of all initially random inactivation can be skewed further during the life of a female by stochastic or selective processes

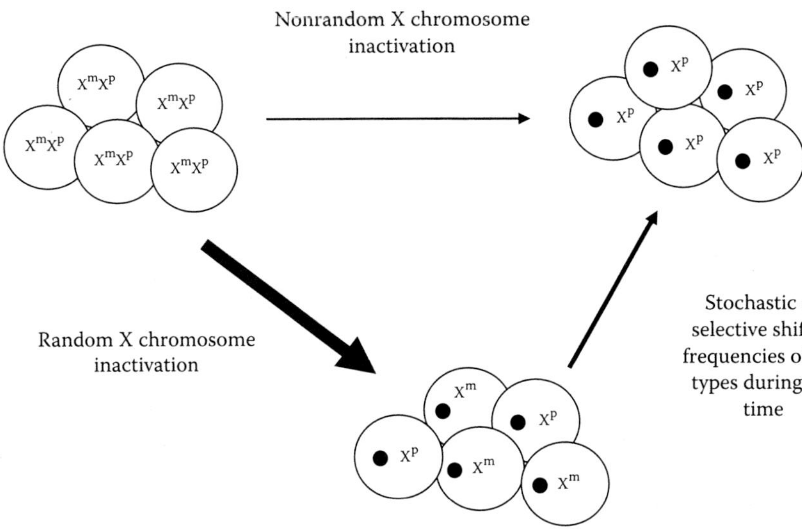

Figure 4.6 Types of X chromosome inactivation in mammalian females. In the preimplantation embryo both paternal and maternal X chromosomes are active (top left). Random inactivation is the prevailing process (represented by a thick arrow) resulting in mosaic individuals carrying more or less equal proportion of cells with inactive maternal or paternal chromosomes (bottom). Inactivated X chromosomes are represented by large black dots. A degree of skewness in distribution of alternative cellular types is expected due to randomness of X chromosome inactivation. Stochastic or selective shift in the frequencies of cell types in some tissues is possible during the life time. This may increase skewness of the distribution. Nonrandom inactivation is rare but possible due to significant differences in *Xic* of maternal or paternal origin. This may lead rarely to extremely skewed distributions (top right). (Redrawn from Minks, Robinson, and Brown 2008, with modifications.)

which may affect frequencies of cellular clones with particular genotype or phenotype. Second, it could be a result of initially nonrandom X chromosome inactivation caused by differences in maternal and paternal *Xce* alleles responsible for the inactivation process. Indeed, several cases were discovered in which the differences in genetic structure of the X inactivation center have caused preferential inactivation of either paternal or maternal chromosome. This is rather typical for hybrids, when the parents originated from distant populations or subspecies (Zakian et al. 1987).

Thus, X chromosome inactivation is a complex process with strict genetic control which evolved in placental mammals in order to resolve the dosage compensation problem. It is most likely that the equal frequency of cells carrying an active copy of the paternal or maternal X chromosome is beneficial for mammalian females. The X chromosome inactivation system has both stochastic and deterministic features (Edwards and Bestor 2007). The existence of a genetically controlled device able to generate randomness in mammalian females is another strong argument in favor of the important role played by random processes in life.

Gene networks and canalization

As we saw earlier in the book, stochastic and random processes have a strong impact on transformation of genetic information into phenotypic. The level of noise in many biological processes is high and constant changes of environmental conditions make this noise even more powerful. Certainly there are margins within which this phenotypic variability is tolerable. Thus, the question is how the intensity of the noise is kept within acceptable limits. This is not a new question. Significant progress in understanding this matter was achieved by C. H. Waddington (1942) and independently by I. I. Shmalgauzen (1949). The idea of canalization of development and the role of stabilizing selection in this process was proposed by these researchers. It was known for a long time that the wild phenotypes, which are typically observed in nature, are much less variable than most mutant forms. The traditional explanation takes into consideration dominance because the majority of phenotype-effecting mutations are recessive and usually can be observed only in homozygotes, which are not highly likely for mutant alleles with low frequency. There are two other factors. First, some genes have duplicates, and potential changes caused by mutation in one gene can be buffered by others. Second, genes and proteins are involved in numerous complex biochemical and developmental interactions, which in some way are responsible for creating homeostasis of development. Stabilizing selection constantly removes from populations phenotypes that deviate significantly from the wild type and have low fitness. This favors developmentally robust genotypes and enhances the homeostasis of developmental processes.

During the last decade similar problems became the focal point for a new round of research efforts, which are based on the latest genomic data, computer modeling, and ideas of genetic and developmental networks (Barabási and Oltvai 2004). The analytical approach applied to molecular biological, biochemical, and genetic systems, as well as to genomes, yielded a great wealth of information, which can be used for building realistic networks models. There is a strong hope that such a synthetic approach will elucidate mechanisms buffering stochastic variations in gene expression and development.

Networks constructed from the same number of components may have very different topologies. Among them hierarchical networks have a set of promising biological characteristics, including modularity, redundancy of connections between nodes, local clustering, and scale-free topology (Barabási and Oltvai 2004). The latter characteristic refers to a structure containing many nodes with a few connections and few nodes with numerous connections. Such connection-reach nodes are called hubs, and within a genetic network they can represent key regulatory genes, proteins, and other centrally positioned signalling molecules. The redundancy of gene or protein connections is the major factor providing robustness to genetic networks against mutations and stochastic noise (Wagner 2000). Robustness to mutations and stochastic fluctuations are correlated in networks despite the rarity of mutations and ubiquity of stochastic fluctuations (Ciliberti et al. 2007). It was found that gene duplications might be responsible for some genetic buffering but they are relatively minor contributors to the process. Genes involved in redundant networks evolved faster (Kitami and Nadeau 2002). It was also shown that "the production of essential and complex-forming proteins involves lower levels of noise than does the production of most other genes" (Fraser et al. 2004). Interestingly genes that are upregulated in development have lower variability, and as these genes are downregulated it increases variability, indicating that developmental context may affect the evolution of gene expression (Rifkin et al. 2005).

Hub proteins, which were defined as those that interact with at least ten counterparts, are characterized by more significant intrinsic structural disorder than "end" proteins that have only one contact (Haynes et al. 2006). A lack of rigid structure in lengthy regions of hub proteins creates sufficient flexibility for interactions with numerous proteins. Existence of proteins with high connectivity seems to be important for gene/protein networks and their robustness to mutational, stochastic, and environmental influences. It was suggested that robustness of networks is an evolvable property and this may be a general organizational principle of biological networks (Ciliberti et al. 2007). Similarly gene intrinsic noise and stochastic developmental variations might be the subject of selection (Fraser et al. 2004; Kærn et al. 2005). Earlier mentioned experiments with selection of nongenetic variability in parthenogenetic clones are in tune with this general conclusion (Ruvinsky et al. 1983). In such cases epigenetic features of dynamic genetic systems might be passed to the next generation (Jablonka and Lamb 1995).

What is known about the origin of genetic networks? So far there is little evidence that the networks were built up by natural selection. Certainly a lack of evidence is not proof that the genetic networks evolved without a considerable influence of natural selection. Nevertheless a suggestion was made that a number of qualitative features of the networks could arise due to nonadaptive processes like genetic drift, mutation, and recombination, raising questions about whether natural selection is necessary or even sufficient for the origin of many aspects of gene-network topologies (Lynch 2007). Similar views are advocated by Ciliberti et al. (2007), who considered possibilities for neutral evolution of genetic networks, particularly evolution of robustness, without negating the role of natural selection in elimination of deleterious mutations. This is a complicated issue because robustness being generally advantageous might not provide immediate selective advantages. Significant efforts are required to remove semantic difficulties from this kind of discussion; even more is necessary to prove that some networks might be built by random processes without selection.

Self-organization of complex systems is a well known, yet underestimated, phenomenon. It usually relies on positive and negative feedback and multiple interactions between the components of the system, duplication of components and processes, and in some cases the hierarchical structure. Genetic and other biological networks fit well into the category of self-organizing systems. Genetic networks have a significant propensity to produce canalization or in other words reduction of stochastic noise on different levels of biological organizations. Canalization may be an inevitable outcome of complex interactions within the networks (Siegal and Bergman 2002). Hub genes and proteins have strong influence on the ability of networks to buffer fluctuations and keep the system close to optimum and thus bring stability. However, when such hub nodes of the networks are under extreme pressure or compromised, this may have a very significant effect on stability of numerous components and the whole system (Siegal, Promislow, and Bergman 2007). For instance highly connected genes or proteins are more likely to have lethal or highly destabilizing effects when mutations drastically change their function.

Compelling evidence of destabilization caused by intensive selection for tame behavior was obtained in silver foxes. This long-term selection of wild animals brought a great deal of transformation in numerous traits, including behavior, reproductive pattern, morphology, developmental rates, physiology, and endocrinology. These massive changes occurred during a few dozen generations, a very short evolutionary interval, and emulated reorganizations typical for dogs and other domestic species (Belyaev 1979; Trut 1999; Trut, Oskina, Kharlamova 2009). Animal behavior is controlled by hormones and neurotransmitters which have a high degree of connectivity within the developmental and gene network and, if so, hold a top hierarchical position in the developmental network. It seems likely that the observed phenotypic changes were caused by changes in a limited number of genes playing key regulatory roles. Even small genetic changes picked up by selection for behavior in these genes "could produce a cascade of changes in gene activity and, as a consequence, frequent and extensive changes in the phenotypes" (Trut 2001).

Types of randomness

The obvious outcome of stochasticity in transferring information from genes to phenotypes is uncertainty of development. It means that the "best genotypes" do not always lead to the "best phenotypes" or more generally any genotype has a multitude of potential realizations. This happens not only due to a constant barrage of environmental factors but also because of loose connections between genotypes and phenotypes caused by several factors, including stochasticity of development. As a consequence the effectiveness of selection is reduced.

However, the noisy gene activities and interactions during development provide some benefits, not only challenges (Kærn et al. 2005). In general terms the main benefit is flexibility of development. The degree of canalization depends on the importance of the traits under consideration. For instance, production of essential proteins might be under tighter control than many others (Fraser et al. 2004). It appears likely that numerous factors affect the degree of developmental canalization but the current understanding is still limited. It was suggested that microRNAs play a role in canalization and thereby in maintaining robustness of development (Hornstein and Shomron 2006). Contrary to the traditional explanation, computer modeling indicates that evolution of robustness is not necessarily caused by stabilizing selection. When the developmental process is modeled as a network of interacting transcription factors it leads to canalization regardless of intensity of stabilizing selection (Siegal and Bergman 2002). Mathematical analysis of high-dimensional nonlinear genetic regulatory networks (GRN) with basic parameters akin to real systems reveals that an inherent property of such GRN is the ability to reside in a state of stationary pseudo-random fluctuations which may be regarded as a stochastic process. This stationary state develops despite the absence of explicit mechanisms for suppressing instability (Rosenfeld 2007). Despite the role that stabilizing selection may or may not play in canalization of development, the major conclusion that robustness of GRN is an evolvable trait is solid (Ciliberti et al. 2007).

Deterministic and stochastic effects in gene regulation and development coexist and this fact is widely accepted. It also seems only natural to consider stochastic variation as a natural by-product of highly complex molecular and cellular systems. As this chapter shows, such a view does not represent the whole truth. There are at least three independent sources of randomness in development. The first one, which we discussed earlier in Chapter 2, has a quantum nature and originates on the subatomic level. Substitutions of nucleotides in DNA vividly exemplify this type of randomness. The exact threshold above which quantum uncertainty loses its direct influence is difficult to identify. It looks likely, however, that this uncertainty affects not only the origin of new mutations but a broader spectrum of molecular phenomena. Apparently this source of randomness is not caused by complexity of molecular or cellular systems.

The second source of randomness is indeed generated by the complexity of molecular, subcellular, and cellular interactions and, even if such interactions were purely deterministic, the only available description is probabilistic. Gene expression noise, numerous epigenetic phenomena on the cellular and organismal levels, stochastic fluctuations in genetic networks, and complex developmental processes likely represent this type of stochasticity. A significant fraction of recombination events also belongs to this category. It might well be that not only the description but the essence of the majority of these processes is probabilistic. Still the border line separating quantum uncertainty and stochastic fluctuations has not been clearly drawn so far.

The third source of randomness is entirely different and cannot be found in nonliving matter. It comprises those processes that were formed in the course of evolution in order to resolve specific biological problems. Among them are very distinct phenomena, such as random inactivation of X chromosomes in mammalian females and random inactivation of some autosomal genes; somatic rearrangements of immune genes and of course segregation of homologous chromosomes during the first meiotic division (see Chapter 5). Existence of this type of randomness demonstrates that biological systems consume huge amounts of additional variation in order to meet the challenges of the surrounding world and to develop suitable adaptations. It is useful to recognize different types of randomness in biological systems; otherwise a proper assessment of their biological significance is intricate.

The first type of randomness, as generally accepted, is the major source of mutations, which are the primary source of genetic variability and essential for evolution. The biological importance of the second type of randomness is still less understood. The majority of epigenetic events described in this chapter are the result of complex stochastic processes and their intensity is under control. There is mounting evidence that in some cases epigenetic changes can be transmitted between generations

(Jablonka and Lamb 1995; Whitelaw and Whitelaw 2006). The latest data confirm that epigenetic changes caused by environmental influences may have transgenerational effects (Rassoulzadegan et al. 2006; Crews et al. 2007). Further efforts are needed to draw clearer conclusions about the role of stochasticity caused by epigenetic changes, particularly in evolution. Perhaps it is fair to say that the range of such fluctuations is broad and the consequences are far reaching.

The third category of stochastic events, while it includes very well known biological phenomena, has not been considered so far as a separate entity. The reason behind this is quite obvious; the phenomena comprising this category are very much different. There is only one uniting feature: all these processes have "purely biological" origin and emerged during evolution as specific devices generating additional randomness.

Summary

In this chapter we considered factors influencing transfer of information from genotypes to phenotypes. Numerous mutations and recombination events, which occur during development, generate a significant "noise" and complicate unravelling genotypic information. The environment is another powerful factor influencing development. The question is whether curbing environmental as well as mutation and recombination influences could shorten and straighten the distance between genotype and phenotype. Certainly this is only a theoretical possibility as in reality none of these influences can be completely avoided. It is possible, however, to minimize such influences. Identical twins in mammals and ameiotic clones in invertebrates traditionally provide an excellent opportunity for investigating developmental noise in multicellular organisms. Results of such studies have been available for a long time and epigenetic events are assumed to be responsible for a large "residual" variability. A potential role of epigenetic factors in generating this kind of variability was recognized in the early 1940s, and since the 1970s modeling, mathematical theory, and experimental studies have been steadily emerging.

Activation and inactivation of genes is the essence of the developmental processes and dependent on epigenetic modifications of genes and entire chromosomes. Epigenetic events being dynamic and stochastic constitute a major source of developmental noise. During the last few years another powerful resource of stochasticity has been identified and measured. This is stochasticity of gene expression, which can be observed in genetically identical cells exposed to the same environmental conditions. Such intrinsic gene-specific noise is related to several aspects of transcription and translation and imposes a fundamental limit on precision of gene expression. The stochastic nature of chemical reactions creates an additional layer of variability in a population of genetically identical cells.

Quantitative deterministic description of these processes is not possible and hence probabilistic description is the only option.

The discovery of DNA methylation and chemical modifications of histones were important steps in understanding the chemical nature of epigenetic variability. Each epigenetic event is basically caused by interactions of DNA with proteins, RNA, and other molecules. Such interactions, even if they have deterministic components, are characterized by stochastic behavior and have low predictability and high uncertainty. They change the fate of cell lines and eventually lead to a variety of phenotypes, some of which are pathological, including certain forms of cancer and other genetic disorders. Amplification of the epigenetic events through cell reproduction is the vehicle for their "fixation" in phenotypes of cells and organisms.

Alternative splicing is another widespread process that boosts cellular diversity, particularly in higher eukaryotes, by increasing variety of mRNAs and proteins produced by the same gene. According to existing estimates 35% to 60% of human genes are involved in alternative splicing. Obviously this kind of "molecular mosaicism" creates immense coding opportunities and plasticity. In many situations alternative splicing is a highly regulated process that leads to production of different isoforms in different cell types or tissues. The functional importance of alternative splicing in such cases is undoubtable. There is, however, much evidence that mutations in exons and probably introns also lead to forms of alternative mRNA splicing that intensify developmental stochasticity and cause drastic phenotypic deviations.

The situation observed during gene expression and development is somewhat similar to that found in statistical mechanics and thermodynamics more than a century ago. A reasonable way of describing the behavior of such very complex systems is introduction of general parameters. Instead of common temperature, pressure, and other parameters used in thermodynamics, newly suggested parameters for gene expression may include potential energy landscapes, energetic barriers, escape times, etc. Computer modeling based on mathematical and physical approaches became helpful in understanding and prediction of complex and "noisy" biological systems. Noise in development is not only unavoidable but also causes consequences. On one hand it may lead to developmental abnormalities and hence diseases like cancer; on the other hand it can broaden the potential for adaptation and be used in evolution.

Canalization or buffering of phenotypic variations during development needs an explanation based on the latest research results. Genetic networks, which have been intensively investigated in recent years, seem to be able to provide such an explanation. So-called hierarchical networks have several promising biological characteristics and among them is dynamic stability and thus an ability to contain excessive stochastic noise.

References

Barabási, A.L., and Z.N. Oltvai. 2004. Network biology: Understanding the cell's functional organization. *Nature Reviews Genetics* 5:101–113.

Becskei, A., B.B. Kaufmann, and A. van Oudenaarden. 2005. Contributions of low molecule number and chromosomal positioning to stochastic gene expression. *Nature Genetics* 37:937–944.

Belyaev, D.K. 1979. Destabilizing selection as a factor in domestication. *Journal of Heredity* 70:301–308.

Bestor, T.H. 2003. Imprinting errors and developmental asymmetry. *Philosophical Transactions of the Royal Society of London. B. Biological Sciences* 358:1411–1415.

Bock, C., and T. Lengauer. 2008. Computational epigenetics. *Bioinformatics* 24:1–10.

Bolduc, V., P. Chagnon, S. Provost, M.P. Dubé, C. Belisle, M. Gingras, L. Mollica, and L. Busque. 2008. No evidence that skewing of X chromosome inactivation patterns is transmitted to offspring in humans. *Journal of Clinical Investigation* 118:333–341.

Boumil, R.M., and J.T. Lee. 2001. Forty years of decoding the silence in X-chromosome inactivation. *Human Molecular Genetics* 10:2225–2232.

Cartegni, L., S.L. Chew, and A.R. Krainer. 2002. Listening to silence and understanding nonsense: Exonic mutations that affect splicing. *Nature Reviews Genetics* 3:285–298.

Celotto, A.M., and B.R. Graveley. 2001. Alternative splicing of the *Drosophila* Dscam pre-mRNA is both temporally and spatially regulated. *Genetics* 159:599–608.

Chen, Z.X., and A.D. Riggs. 2005. Maintenance and regulation of DNA methylation patterns in mammals. *Biochemistry and Cell Biology* 83(4):438–448.

Chern, T.-M., E. van Nimwegen, C. Kai, J. Kawai, P. Carninci, Y. Hayashizaki, and M. Zavolan. 2006. A simple physical model predicts small exon length variations. *PLoS Genetics* 2:606–613.

Chess, A., I. Simon, H. Cedar, and R. Axel. 1994. Allelic inactivation regulates olfactory receptor gene expression. *Cell* 78:823–834.

Ciliberti, S., O.C. Martin, and A. Wagner. 2007. Robustness can evolve gradually in complex regulatory gene networks with varying topology. *PLoS Computational Biology* 3:164–173.

Crews, D., A.C. Gore, T.S. Hsu, N.L. Dangleben, M. Spinetta, T. Schallert, M.D. Anway, and M.K. Skinner. 2007. Transgenerational epigenetic imprints on mate preference. *Proceedings of the National Academy of Sciences USA* 104:5942–5946.

Dodd, I.B., M.A. Micheelsen, K. Sneppen, and G. Thon. 2007. Theoretical analysis of epigenetic cell memory by nucleosome modification. *Cell* 129:813–822.

Edwards, J.R., and T.H. Bestor. 2007. Stochastic and deterministic effects in gene regulation. *Heredity* 99:243–244.

Eggan, K., K. Baldwin, M. Tackett, J. Osborne, J. Gogos, A. Chess, R. Axel, and R. Jaenisch. 2004. Mice cloned from olfactory sensory neurons. *Nature* 428:44–49.

Elowitz, M.B., A.J. Levine, E.D. Siggia, and P.S. Swain. 2002. Stochastic gene expression in single cell. *Science* 297:1183–1186.

Falconer, D.S., and P.J. Avery. 1978. Variability of chimeras and mosaics. *Journal of Embryology and Experimental Morphology* 43:195–215.

Flanagan, J.M., V. Popendikyte, N. Pozdniakovaite, M. Sobolev, A. Assadzadeh, A. Schumacher, M. Zangeneh, L. Lau, C. Virtanen, S.-C. Wang, and A. Petronis. 2006. Intra- and interindividual epigenetic variation in human germ cells. *American Journal of Human Genetics* 791:67–84.

Flood, W.D., and A. Ruvinsky. 2000. Alternative splicing and expressivity of the *Axin*^{Fu} allele in mice. *Heredity* 87:146–152.

Fraga, M.F., E. Ballestar, M.F. Paz, S. Ropero, F. Setien, M.L. Ballestar, D. Heine-Suñer, J.C. Cigudosa, M. Urioste, J. Benitez, M. Boix-Chornet, A. Sanchez-Aguilera, S. Ling, E. Carlsson, P. Poulsen, A. Vaag, Z. Stephan, T.D. Spector, Y.-Z. Wu, C. Christoph Plass, and M. Esteller. 2005. Epigenetic differences arise during the lifetime of monozygotic twins. *Proceedings of the National Academy of Sciences USA* 102:10604–10609.

Fraser, H.B., A.E. Hirsh, G. Giaever, J. Kumm, and M.B. Eisen. 2004. Noise minimisation in eukaryotic gene expression. *PLoS Biology* 2:834–838.

Gimelbrant, A., J.N. Hutchinson, B.R. Thompson, and A. Chess. 2007. Widespread monoallelic expression on human autosomes. *Science* 318:1136–1140.

Gottlieb, B., L.K. Beitel, and M.A. Trifiro. 2001. Somatic mosaicism and variable expressivity. *Trends in Genetics* 17:79–83.

Haynes, C., C.J. Oldfield, F. Ji, N. Klitgord, M.E. Cusick, P. Radivojac, V.N. Uversky, M. Vidal, and L.M. Yakoucheva. 2006. Intrinsic disorder is a common feature of hub proteins from four eukaryotic interactomes. *PLoS Computational Biology* 2:890–901.

Healey, S.C., K.M. Kirk, V.J. Hyland, C.F. Munns, A.K. Henders, J.A. Batch, A.C. Heath, N.G. Martin, and I.A. Glass. 2001. Height discordance in monozygotic females is not attributable to discordant inactivation of X-linked stature determining genes. *Twin Research* 4:19–24.

Hoffmann, M.J., and W.A. Schulz. 2005. Causes and consequences of DNA hypomethylation in human cancer. *Biochemistry and Cell Biology* 83(3):296–321.

Holliday, R. 1990. Mechanisms for the control of gene activity during development. *Biological Reviews of the Cambridge Philosophical Society* 65:431–471.

Holliday, R. 2005. DNA methylation and epigenotypes. *Biochemistry* (Moscow). 70:500–504.

Holliday, R., and J.E. Pugh. 1975. DNA modification mechanisms and gene activity during development. *Science* 187:226–232.

Hornstein, E., and N. Shomron. 2006. Canalization of development by microRNAs. *Nature Genetics* 38:S20–S24.

Humpherys, D., K. Eggan, H. Akutsu, K Hochedlinger, W.M. Rideout III, D. Biniszkiewicz, R. Yanagimachi, and R. Jaenisch. 2001. Epigenetic instability in ES cells and cloned mice. *Science* 293:95–97.

Jablonka, E., and M.J. Lamb. 1995. *Epigenetic inheritance and evolution*. New York: Oxford University Press.

Kærn, M., T.C. Elston, W.J. Blake, and J.J. Collins. 2005. Stochasticity in gene expression: From theories to phenotypes. *Nature Reviews Genetics* 6:451–464.

Karmakar, R., and I. Bose. 2004. Graded and binary responses in stochastic gene expression. *Physical Biology* 1:197–204.

Kazazian, H.H., Jr. 2004. Mobile elements: Drivers of genome evolution. *Science* 303:1626–1632.

Kitami, T., and J.H. Nadeau. 2002. Biochemical networking contributes more to genetic buffering in human and mouse metabolic pathways than does gene duplication. *Nature Genetics* 32:191–194.

Ko, M.S. 1992. Induction mechanism of a single gene molecule: Stochastic or deterministic? *Bioessays* 14:341–346.

Kussell, E., and S. Leibler. 2005. Phenotypic diversity, population growth, and information in fluctuating environments. *Science* 309:2005–2007.

Le Douarian, N., and A. McLaren, eds. 1984. *Chimeras in developmental biology*. London: Academic Press.

Levine, E., and T. Hwa. 2007. Stochastic fluctuations in metabolic pathways. *Proceedings of the National Academy of Sciences USA* 104:9224–9229.

Lynch, M. 2007. The evolution of genetic networks by non-adaptive processes. *Nature Reviews Genetics* 10:803–811.

Martin, G.M. 2005. Epigenetic drift in aging identical twins. *Proceedings of the National Academy of Sciences USA* 102:10413–10414.

Martin, D.I., R. Ward, and C.M. Suter. 2005. Germline epimutation: A basis for epigenetic disease in humans. *Annals of the New York Academy of Sciences* 1054:68–77.

McClintock, B. 1953. Induction of instability at selected loci in maize. *Genetics* 38:579–599.

McLaren, A. 1995. Germ cells and germ cell sex. *Philosophical Transactions of the Royal Society of London. B. Biological Sciences* 350:229–233.

Minks, J., W.P. Robinson, and C.J. Brown. 2008. A skewed view of X chromosome inactivation. *Journal of Clinical Investigation* 118:20–23.

Monkhorst, K., I. Jonkers, E. Rentmeester, F. Grosveld, and J. Gribnau. 2008. X inactivation counting and choice is a stochastic process: Evidence for involvement of an X-linked activator. *Cell* 132:410–421.

Nitasaka, E., and Yamazaki, T. 1991. A novel eye morphology induced by a P element in somatic tissue of *Drosophila melanogaster*. *Molecular and General Genetics* 227:181–186.

Ohlsson, R., A. Paldi, and J.A. Graves. 2001. Did genomic imprinting and X chromosome inactivation arise from stochastic expression? *Trends in Genetics* 17:136–141.

Olson, S., M. Blanchette, J. Park, Y. Savva, G.W. Yeo, J.M. Yeakley, D.C. Rio, B.R. Graveley. 2007. A regulator of Dscam mutually exclusive splicing fidelity. *Nature Structural and Molecular Biology* 14:1134–1140.

Paixão, T., T.P. Carvalho, D.P. Calado, and J. Carneiro. 2007. Quantitative insights into stochastic monoallelic expression of cytokine genes. *Immunology and Cell Biology* 85:315–322.

Petronis, A. 2006. Epigenetics and twins: Three variations on the theme. *Trends in Genetics* 22:347–350.

Petronis, A., I.I. Gottesman, P. Kan, J. Kennedy, A.D. Paterson, and V. Popendikyte. 2003. Monozygotic twins exhibit numerous epigenetic differences: Clues to twin discordance? *Schizophrenia Bulletin* 29:169–178.

Raser, J.M., and E.K. O'Shea. 2004. Control of stochasticity in eukaryotic gene expression. *Science* 304:1811–1814.

Rassoulzadegan, M., V. Grandjean, P. Gounon, S. Vincent, I. Gillot, and F. Cuzin. 2006. RNA-mediated non-mendelian inheritance of an epigenetic change in the mouse. *Nature* 441:469–474.

Redina, O.E., A.I. Zhelezova, A.N. Golubitsa, A. Agulnik, and A. Ruvinsky. 1994. Phenotypic expression of the *fused (Fu)* gene in chimaeric mice. *Genetics Research* 63:183–187.

Rifkin, S.A., D. Houle, J. Kim, and K.P. White. 2005. A mutation accumulation assay reveals a broad capacity for rapid evolution of gene expression. *Nature* 438:220–223.

Romero, P.R., S. Zaidi, Y.Y. Fang, V.N. Uversky, P. Radivojac, C.J. Oldfield, M.S. Cortese, M. Sickmeier, T. LeGall, Z. Obradovic, and A.K. Dunker. 2006. Alternative splicing in concert with protein intrinsic disorder enables increased functional diversity in multicellular organisms. *Proceedings of the National Academy of Sciences USA* 103:8390–8395.

Rosenfeld, S. 2007. Stochastic cooperativity in non-linear dynamics of genetic regulatory networks. *Mathematical Biosciences* 210:121–142.

Ruvinsky, A., W.D. Flood, T. Zhang, and F. Costantini. 2000. Unusual inheritance of the *Axin*^{Fu} mutation in mice is associated with widespread rearrangements in the proximal region of chromosome 17. *Genetics Research* 76:135–147.

Ruvinsky, A.O., Yu.I. Lobkov, and D.K. Belyeav. 1983. Spontaneous and induced activation of genes affecting the phenotypic expression of glucose 6-phosphate dehydrogenase in *Daphnia pulex.* I. Intraclonal variations in the electrophoretic mobility of G6PD. *Molecular and General Genetics* 189:485–489.

Sano, Y., T. Shimada, H. Nakashima, R.H. Nicholson, J.F. Eliason, T.A. Kocarek, and M.S. Ko. 2001. Random monoallelic expression of three genes clustered within 60 kb of mouse t complex genomic DNA. *Genome Research* 11:1833–1841.

Shmalgauzen, I.I. 1949. *Factors of evolution: The theory of stabilizing selection*, trans. I. Dordick. Philadelphia: Blackiston.

Siegal, M.L., and A. Bergman. 2002. Waddington canalization revisited: Developmental stability and evolution. *Proceedings of the National Academy of Sciences USA* 99:10528–10532.

Siegal, M.L., D.E. Promislow, and A. Bergman. 2007. Functional and evolutionary inference in gene networks: Does topology matter? *Genetica* 129:83–103.

Tchuraev, R.N. 2006. Epigenetics: Gene and epigene networks in ontogeny and phylogeny. *Russian Journal of Genetics* 42:1066–1083.

Timoféeff-Ressovsky, H.A., and N.W. Timoféeff-Ressovsky. 1926. Über das phänotypische Manifestation des Genotypes. II. Über idio-somatische Variationsgruppen bei Drosophila funebris. *Wilhelm Roux' Archiv für Entwicklungsmechanik der Organismen* 108:146–170.

Ting, A.H., K.M. McGarvey, and S.B. Baylin. 2006. The cancer epigenome: Components and functional correlates. *Genes and Development* 20(23):3215–3231.

Tress, M.L., P.L. Martelli, A. Frankish, G.A. Reeves., J.J. Wesselink, C. Yeats, P.L. Olason, M. Albrecht, H. Hegyi, A. Giorgetti, D. Raimondo, J. Lagarde, R.A. Laskowski, G. López, M.L. Sadowski, J.D. Watson, P. Fariselli, I. Rossi, A. Nagy, W. Kai, Z. Størling, M. Orsini, Y. Assenov, H. Blankenburg, C. Huthmacher, F. Ramírez, A. Schlicker, F. Denoeud, P. Jones, S. Kerrien, S. Orchard, S.E. Antonarakis, A. Reymond, E. Birney, S. Brunak, R. Casadio, R. Guigo, J. Harrow, H. Hermjakob, D.T. Jones, T. Lengauer, C.A. Orengo, L. Patthy, J.M. Thornton, A. Tramontano, and A. Valencia. 2007. The implications of alternative splicing in the ENCODE protein complement. *Proceedings of the National Academy of Sciences USA* 104:5495–5500.

Trut, L.N. 1999. Early canid domestication: Farm-fox experiment. *American Scientist* 87:160–169.

Trut, L.N. 2001. Experimental studies of early canid domestication. In *The genetics of the dog*, ed. A. Ruvinsky and J. Sampson. Wallingford, UK: CABI Publishing.

Trut, L., I. Oskina, and A. Kharlamova. 2009. Animal evolution during domestication: the domesticated fox as a model. *BioEssays* 31:349–360.

Tsujita, T., N. Niikawa, H. Yamashita, A. Imamura, A. Hamada, Y. Nakane, and Y. Okazaki. 1998. Genomic discordance between monozygotic twins discordant for schizophrenia. *American Journal of Psychiatry* 155:422–424.

Veitia, R.A. 2005. Stochasticity or the fatal "imperfection" of cloning. *Journal of Biosciences* 30:21–30.

Verger, A., and M. Crossley. 2005. The transcriptome. In *Mammalian genomics*, ed. A. Ruvinsky and J.A. Marshall Graves, 117–151. Wallingford, UK: CABI Publishing.

Vogt, G., M. Huber, M. Thiemann, G. van der Boogaart, O.J. Schmitz, and C.D. Schubart. 2008. Production of different phenotypes from the same genotype in the same environment by developmental variation. *Journal of Experimental Biology* 211:510–523.

Waddington, C.H. 1942. Canalization of development and the inheritance of acquired characters. *Nature* 150:563–565.

Wagner, A. 2000. Robustness against mutations in genetic networks of yeast. *Nature Genetics* 24:355–361.

Whitelaw, E., and D. Garrick. 2005. The epigenome. In *Mammalian genomics*, ed. A. Ruvinsky and J.A. Marshall Graves, 179–200. Wallingford, UK: CABI Publishing.

Whitelaw, N.C., and E. Whitelaw. 2006. How lifetimes shape epigenotype within and across generations. *Human Molecular Genetics* 15 (Review Issue 2):R131–R137.

Yochem, J., and R.K. Herman. 2003. Investigating *C. elegans* development through mosaic analysis. *Development* 130:4761–4768.

Zakian, S.M., N.A. Kulbakina, M.N. Meyer, L.A. Semenova, M.N. Bochkarev, S.I. Radjabli, O.L. Serov. 1987. Non-random inactivation of the X-chromosome in interspecific hybrid voles. *Genetics Research* 50:23–27.

Zhuravel, D., and M. Kærn. 2005. Physics takes another stab at biological design principles. *Molecular Systems Biology* 1:2005.0029.

chapter five

Organized randomness

> It remains, therefore, purely a matter of chance which of the two sorts of pollen will become united with each separate egg cell.
>
> **Gregor Mendel**
> *Experiments in Plant Hybridization (1865)*

Gregor Mendel's vision

The concept of equal segregation is central to the genetics of eukaryotes. Gregor Mendel developed this idea in the mid-nineteenth century using the results of garden pea crosses. According to this principle the members of a gene pair or, using standard terminology, alleles have equal chance to enter a gamete. It means that approximately 50% of gametes produced by a heterozygote Aa, will carry allele A and the rest will carry allele a. Mendel's discovery became widely known at the beginning of the twentieth century and has been a subject of fascination ever since. Regardless of what Mendel's preconceived ideas were, one thing is clear: he found a highly organized source of randomness. As was confirmed later this source of randomness indeed exists in eukaryotic species with meiotic production of gametes. At the time Mendel wrote his major papers nothing was known about meiosis, chromosomes, and genes. Meiosis was discovered by the German biologist Oscar Hertwig in 1876, ten years after the publication of Mendel's major work. The term *chromosome* was introduced in 1888 by another German biologist Heinrich von Waldeyer. The role of meiosis was first recognized by the outstanding German biologist August Weismann in 1890, and his contribution to the development of genetics and the theory of evolution was exceptional. The formulation of chromosome theory, which connected chromosome behavior in meiosis and the formation and fertilization of gametes, was achieved independently by Theodor Boveri in Germany and Walter Sutton in the United States in 1902 (Box 5.1 and Box 5.2; see Figure 5.1 and Figure 5.2). Finally starting from 1911 Thomas Morgan (see his biography in Chapter 3, Box 3.1) developed the chromosome theory of heredity which fused together cytological and genetic data. Interestingly enough the first reaction of Morgan to

BOX 5.1 THEODOR BOVERI

Theodor Boveri was born in 1862 in Bamberg, Germany. In 1881 he entered the University of Munich as a student of anatomy and biology. He received his Ph.D. in 1885 and was fortunate to continue research at the University of Munich in the laboratory of Richard Hertwig, who a decade earlier discovered meiosis and fertilization in sea urchin. In 1891, Hertwig invited Boveri to take a position of assistant professor. Two years later Boveri was appointed professor of zoology and comparative anatomy at the University of Würzburg and held this chair until he died on 1915. Studying meiosis and early cleavage in horse nematodes he realized that despite numerous morphological changes individual chromosomes maintain the integrity of their structures during cell cycle and division. It was the hypothesis of chromosome individuality (1887), which built his reputation as a prominent cell researcher.

Boveri began his work with sea urchins during several visits to Naples Zoological Station in Italy. He had shown that insemination of nucleus-free sea urchin eggs is sufficient for normal development. Similar results were obtained when the eggs contained only female chromosomes. Boveri came to the conclusion that sperm and egg contribute an equal number of chromosomes to the zygote. The idea that chromosomes could be the carriers of heredity became quite pertinent. These experiments and conclusions were made not long before the rediscovery of Mendel's work in 1900. Boveri was well prepared to make a decisive step and logically connect the behavior of chromosomes and Mendelian segregation. In publications of 1902 to 1904 he proposed the chromosome theory of heredity. In the same time and independently, Walter Sutton, a graduate student from Columbia University, drew a similar conclusion.

This great scientific breakthrough steadily became the cornerstone of modern biology. Two very influential biological textbooks of the first half of the twentieth century (*The Cell in Development and Heredity* by E.B. Wilson in the United States and *Experimentelle Beiträge zu einer Theorie der Entwicklung* by H. Spemann in Germany) were dedicated to Boveri. In 1918 Edward Wilson wrote: "Boveri stood without a rival among the biologists of his generation; and his writings will long endure as classical models."

Source: http://www.biozentrum.uni-wuerzburg.de/index.php?id=thepersontheodor

BOX 5.2 WALTER S. SUTTON

Walter Stanborough Sutton was born in Utica, New York, in 1877. About ten years later the family moved to Kansas. In 1896, he enrolled at the University of Kansas in engineering, a subject he loved. However, the death of a younger brother affected him profoundly. Soon after this tragic event Sutton was enrolled in biological sciences in preparation for a medical career. He received a bachelor's degree in 1900 and a master's degree in 1901. His master's thesis was devoted to investigation of spermatogenesis in a newly discovered grasshopper species. Sutton began the graduate program at Columbia University with leading American cytologist E.B. Wilson. There in 1902 he wrote a paper which had the following statement: "I may finally call attention to the probability that the association of paternal and maternal chromosomes in pairs and their subsequent separation during the reducing division … may constitute the physical basis of the Mendelian law of heredity" (cited in Crow and Crow 2002; this paper provides valuable information about the time, people, and the discovery). From 1903 to 1905 Sutton worked as a foreman in the Kansas oil fields and then he continued medical studies at the Columbia College of Physicians and Surgeons. Between 1907 and 1909 he accomplished a fellowship in surgery. In 1909 Sutton was appointed assistant professor of surgery at the University of Kansas. He did not finish his Ph.D. in biology and published only two student papers, which connected the meiotic behavior of chromosomes with Mendelian laws and had an enormous influence on the development of genetics and biology in general.

Entirely independently and at the same time the German biologist Theodor Boveri came to similar conclusions. There was no immediate and complete acceptance of these views; however, some years later the Boveri–Sutton chromosome theory became a classical part of genetics. Sutton had a successful career as a surgeon until 1916 when he died suddenly in Kansas City.

Source: Crow, E.F., and J.F. Crow. 2002. 100 years ago: Walter Sutton and the chromosome theory of heredity. *Genetics* 160: 1–4.

the ideas expressed by Boveri and Sutton was not too positive. It took a decade of hard work and thinking before he embraced a similar vision and moved ahead.

The regularity of segregation of homologous chromosomes during the first meiotic division is amazing assuming possible differences between the chromosomes and their allelic compositions. Equal segregation of

Figure 5.1 Theodor Boveri (1862–1915), outstanding German biologist who was a co-author of the chromosome theory in 1902. He was born in Bamberg and died in Würzburg, Germany. (Courtesy of the University of Wűrzburg Library.)

alleles during meiosis, however, is not sufficient to guarantee equal transmission of alleles to the next generation. Fertilization or joining of maternal and paternal gametes must also be random in order to make the probability of allele *a* reaching the next generation the same as for allele *A*. Quite obviously assortative matings and preferential fertilization distort equal allele transmission to the next generation.

Equal transmission of alleles is a very consistent process. The number of known deviations from this rule is rather small despite experimental data accumulated for more than a century. Interestingly the principle of equal transmission of alleles itself was never criticized, while Mendel himself was. Ronald Fisher, the leading English statistician and geneticist, over the years came to the conclusion that the transmission data of some experiments published by Mendel "are too good to be true." Fisher's initial paper, which appeared in 1936, caused a long debate which continues

Figure 5.2 Walter Sutton (1877–1916), famous American biologist and surgeon, a co-author of the chromosome theory in 1902. He was born in New York City and died in Kansas City. (Courtesy of *Genetics*, the journal of the Genetics Society of America.)

today (Weiling 1986; Orel 1996; Hartl and Fairbanks 2007). The main tenet of Fisher's original claim was that some of Mendel's results were too close to expectation, which is very unlikely from a statistical point of view. Here we are not going to consider the arguments of this discussion again because this has been already done perfectly well. The major conclusion of this long saga is absolutely certain; if some of Mendel's published results do not look likely from a statistical point of view it does not mean that the data were tampered with. Simple and convincing explanations have been suggested by several well-respected geneticists, including the outstanding American geneticist and statistician Sewall Wright.

Fisher's comments, right or wrong, do not affect the major conclusion drawn by Mendel and generations of geneticists during more than a century of intensive research (Box 5.3). Nevertheless, the question of why the

BOX 5.3 MENDELIAN LAWS AND IDEOLOGICAL STRUGGLE

Pious Gregor Mendel could not imagine that his discovery will be the subject of discussions in a highly charged political atmosphere. But this is exactly what happened in the USSR from the 1930s to the 1950s. In keeping with the scenario of a classical tragedy, the forces of good and evil collided in an unequal battle. T.D. Lysenko and his team, supported by Stalin, tried to discredit and ban genetics in the USSR. Nikolai I. Vavilov, an outstanding biologist, and other honest Soviet geneticists did their best to resist, preferring to face death rather than betray scientific truth. This resistance cost many innocent lives. Vavilov himself was imprisoned and died of malnutrition in 1943. A brief citation from Vladimir M. Tihomirov's paper published in 2003 and devoted to his teacher, and one of the great mathematicians of the twentieth century, Andrei N. Kolmogorov, provides some insight into those years:

In the year 1940, Kolmogorov published his monograph entitled *A New Confirmation of Laws of Mendel*. During that year a discussion unfolded between geneticists and the followers of Lisenko, concerning the validity of Mendel's laws. In order to settle the dispute, Lisenko (*T.D. Lysenko, AR*) and N.M. Vavilov (*N.I. Vavilov, AR)* have asked their collaborators, N.I. Ermolaev and T.K. Enin, respectively, to replicate Mendel's experiments in order to either disprove (this job was given to Ermolaev), or to prove (the job was given to Enin), Mendel's theory. Both investigators "coped well" with their assignments. Then, Andrei Nikolaevich (*Kolmogorov, AR*) has studied resolutely results obtained by both investigators. In the introduction of the mentioned monograph, he states: Not only that Mendel's theory leads to the simplest conclusion about the stable approximation of the ratio's value 3:1, but gives also a way for predicting the mean deviation from this sample ratio. Based on this statistical analysis of deviations from the ratio 3:1, one gets a new, sharper and more exhaustive method for verification of Mendel's statement about splintering of heredity signs. The goal of the present note is to point out the most rational method for testing and illustrations, which according to the opinion of the author (*A.N. Kolmogorov, AR*), is contained in data provided by N.I. Ermolaev. These data, contrary to what N.M. Ermolaev wishes, provide a new, bright confirmation of Mendel's laws. (Andrei Nikolaevich said that Ermolaev conducted his experiments and data gathering in a remarkably conscientious manner, which enabled him to provide a new, bright confirmation of Mendel's laws, while Enin, who right away wanted to confirm the laws, obviously rejected some results of experiments (which didn't look favorable towards this goal), and by so doing have obtained the better results. About this fudging of Enin, Andrei Nikolaevich states in the monograph: "attaching systematically excessive approximation of frequency m/n to 3/4, is what one finds in data of Enin".

Source: V.M. Tihomirov. 2003. Andrei Nikolaevich Kolmogorov (1903–1987). The great Russian scientist. *The Teaching of Mathematics* 6(1):25–36.

observed segregation ratios are usually close to Mendelian expectation, and why statistically significant deviations are so rare awaits explanations and we shall consider this issue later.

Random segregation, uncertainty, and combinatorial variability

According to Mendel's principle each allele of a heterozygote has ~50% likelihood of getting into a gamete. What is the degree of uncertainty when the chance for either of two alleles (*A* or *a*) to enter a gamete is 50%? Shannon's famous formula (Equation 5.1) for uncertainty allows finding the answer.

$$H = -\sum_{i=1}^{n} p_i \log_2 p_i \tag{5.1}$$

In this formula H stands for uncertainty and p_i is the probability of the ith outcome. In the model, which we consider here, there are only two possibilities: either allele *A* or allele *a* enters a gamete. Assuming $p = 0.5$, it can be easily found that $H = 1$, which is the highest possible value because H varies between 0 and 1. This formal conclusion would match an intuitive expectation that uncertainty is the highest when probability of either of two alleles to be passed from a heterozygote to a gamete is 0.5. Figure 5.3 provides visual support for this conclusion. Uncertainty rapidly declines in both directions from 0.5.

Thus, uncertainty reaches a maximum when $p = 0.5$. From the physical point of view the highest uncertainty is the most stable state as it does not require additional energy for maintaining the status quo. However, from the biological point of view this might not necessarily appear to be so and the question of why Mendelian segregation is the rule rather than the exception was raised many times. This problem is discussed below, in the section "Why Is the First Mendelian Law So Common?"

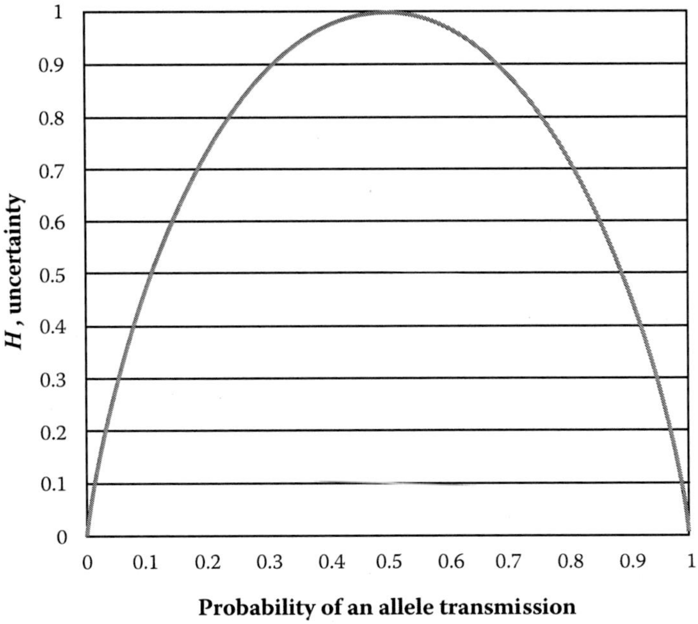

Figure 5.3 Graph of Shannon's *H* function for two alternative outcomes; in this particular case transmission of one of two alleles from a heterozygote to the next generation. Uncertainty reaches the highest value of 1, when the probability of transmission of either allele from a heterozygote to the next generation is 0.5. (Redrawn with modifications from Schneider 2005.)

What are the consequences of random segregation of chromosomes? A very important one is the creation of genetic variability on a gigantic scale. Let us consider, for example, how many variants of gametes a human individual can potentially generate. There are twenty-three pairs of homologous chromosomes, which segregate independently. For simplicity we ignore the fact that pairs of sex chromosomes in males and females are not the same. Assuming that each pair of homologous chromosomes differs at least by one mutation, we come to an estimate that 2^{23} different types of gametes potentially can be produced by an individual. As a pair of homologous chromosomes in reality may differ by hundreds of mutations, a very large number of different recombinant chromosomes can be created by crossing-over. An assumption that there is at least one exchange for each pair of homologous chromosomes will dramatically increase the initial estimate. In reality more than one crossing-over event occurs between the majority of homologous chromosomes. Another important point to consider is the constantly changing location of crossing-over along a chromosome pair. It is difficult to calculate the exact number of possible gamete types that can be produced by an individual, but most likely an estimate is higher than $2^{100} \approx 10^{30}$. Males obviously produce much more gametes than females; in humans about $(8.5 \pm 1.3) \times 10^6$ sperm cells are produced daily per gram of testicular parenchyma (Johnson, Petty, and Neaves 1980). Assuming that the male's total daily production is ~10^8 to 10^9 gametes, the total number of gametes produced by an individual over the course of fifty years can be estimated as ~10^{12} to 10^{13}. Clearly this is a minuscule fraction of the potential number of gametes that can be produced by an individual. Random combination of male and female gametes during fertilization increases the number of potential zygotes produced by a couple even further $(10^{30})^2 = 10^{60}$. Due to constantly occurring mutations and other factors this number has to be even higher. It means that the probability of two individuals having the same genotype is for all practical purposes not different from 0. This is the reason that all human beings are different, with the exception of identical twins. The emergence of multicellular organisms with meiotic production of gametes brought dramatic changes to evolution of life due to tremendously enhanced combinatorial variability and accelerated pace of evolution.

Genes and chromosomes that violate the law

The concept of equal segregation is commonly known in genetics as the first Mendelian law. As mentioned earlier this law is fulfilled in the majority of situations; nevertheless, there are cases when selfish alleles or chromosomes ignore the law, cheat the meiotic partners and significantly increase their chances to get into the next generation. Male and female meiosis despite basic similarities have significant differences. The major difference is the asymmetric nature of female meiosis. Both the first and the second meiotic divisions in females produce two very different cells. One of them, the oocyte, is the larger cell and the second, the polar body, is the smaller cell. Polar bodies disappear, whereas oocytes carry "the torch of life." For chromosomes and genes, entering the oocyte or the polar body represents a drastic difference. Males, on the contrary, produce sperm cells that are equal in size and very active participants in the fertilization process, and selection of faster and more successful sperm cells is an ongoing process.

The differences in the outcomes of male and female gametogenesis lead to two types of deviations from Mendel's first law. These are transmission ratio distortion (TRD) and meiotic drive. TRD is a male-related phenomenon caused by either an impaired formation or function of some sperm cells. Just a few strong cases of TRD have been discovered so far, including *Sd* in *Drosophila melanogaster* (Sandler and Novitski 1957) and t-haplotypes in *Mus musculus* (Lewontin 1968; Lyon 2003). Deviation from equal transmission in both these cases is likely caused by different fertilizing capacity of sperm cells carrying alternative alleles of a particular gene (Seitz and Bennett 1985; Schimenti 2000; Kusano et al. 2003). In the mouse the presence of chromosome inversions seems to be very important for maintaining the TRD system. An inversion is a structural change in which a section of a chromosome is turned around 180 degrees. The classical consequence of heterozygosity for an inversion is a lack of meiotic recombination due to low viability of recombinant gametes.

The first TRD system studied in mice is caused by so-called t-haplotypes, found in wild populations more than seventy-five years ago. Figure 5.4 shows the essential features of the complex structure of t-haplotypes. The presence of four overlapping inversions is the particularly important feature because it disallows recombination and protects allelic composition of t-haplotypes. This makes possible independent evolution of the haplotypes carrying lengthy complexes of genes in both wild and t-chromosomes. Heterozygous t/+ males, if they carry complete t-haplotypes, manage to pass them to nearly 99% of progeny. This is the most extreme deviation from the equal transmission found in any system. Incomplete t-haplotypes show a whole set of very different TRD, from 20% to 80% (Lyon 2003). It was suggested that so-called *Distorter* genes, which operate in an additive manner, have harmful effects on sperm function. Another suspected participant of the process *Responder*, most likely has more sensitive alleles located in wild chromosomes, which do not carry t-haplotypes. As a result sperm cells without t-haplotype are comparatively less mobile and less capable of successfully passing through all stages of fertilization. Intensive efforts to identify major genes involved in the TRD have brought some progress (Schimenti 2000; Bauer et al. 2007).

The TRD advantage of t-haplotypes is hindered by the lethality of t/t homozygotes and other traits affecting fitness. On the population level this leads to a classical balanced polymorphism, when a particular

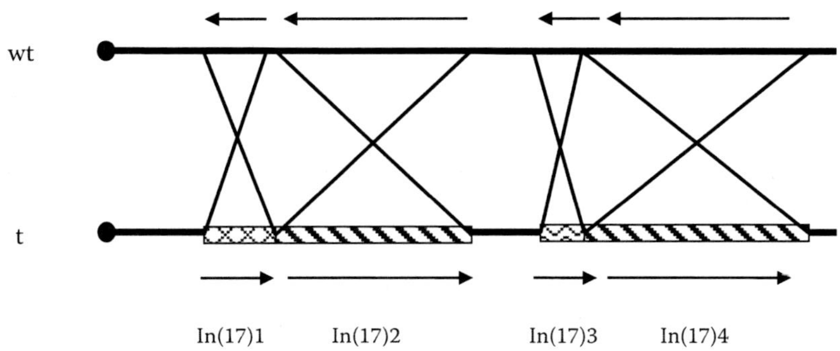

Figure 5.4 Diagram of wild-type (wt)- and t-haplotype (t)-bearing chromosome 17 in the mouse. The inversions are shown as hatched boxes on the t chromosome and indicated as In(17)1, etc. The arrows indicate direction of the DNA segments. (Redrawn with simplifications from Lyon 2003.)

population frequency of t-haplotypes causes a minimal negative impact on the population fitness. This is the equilibrium point, which remains unchanged as long as selective pressures in the population do not change. Despite very aggressive TRD for some naturally occurring t-haplotypes their population frequency usually is not too high, ranging between 7% and 21% (Ruvinsky et al. 1991).

Meiotic drive, on the contrary, is found only in females and can come about by the preferential entry of certain alleles/chromosomes into a polar body or an oocyte at the first or the second meiotic division. This phenomenon is referred to as meiotic drive because the meiotic process is the driving force. A few well-documented cases of meiotic drive have been discovered so far (Ruvinsky 1995; Pardo-Manuel de Villena et al. 2000). Figure 5.5 illustrates the basic facts of meiotic drive of chromosome 1 with the large insertion (named HSR) found in wild populations of mice (Agulnik, Agulnik, and Ruvinsky 1990). There are two meiotic options for the heterozygous females carrying the insertion. The first one, shown on the right side of Figure 5.5 by a dashed thick arrow, assumes that a crossing-over event does occur somewhere between the centromere and the insertion prior to the first meiotic division. As the distance is significant, the likelihood of meiotic recombination is high, which means that this is the most typical scenario. After the crossing-over each chromosome has two chromatids, one with the insertion and another without. The homologous chromosomes look indistinguishable and segregate equally into oocyte and the first polar body, which disappears later. The second meiotic division is different; there is an obvious distinction between the long chromosome with the insertion and the normal chromosome. The thick arrow shows that the most likely outcome of the second meiotic division is retaining the long chromosome in the oocyte.

In the second scenario, shown on the left side of the figure, there is no recombination between the centromere and the insertion. If so, two structurally different chromosomes enter the first meiotic division and the most likely outcome (shown by the thick arrow) is retaining the long chromosome in the oocyte. Thus, meiotic drive in favor of the long chromosome with insertion occurs regardless of the scenario. The intensity of this meiotic drive is quite substantial, about 70%, which means that there is a high probability of the chromosome with the insertion being transmitted from heterozygous females to the next generation. As mentioned earlier the inserted region of chromosome 1 contains the lengthy inversion, which prevents recombination in the region. This is similar to what is observed in t-haplotypes. However, the similarities do not end here. This aberrant chromosome 1 has segregation advantages as well as some disadvantages, among them low fertility and viability of homozygotes. Again, as in the case of t-haplotypes, a balanced polymorphism exists in mice populations, which maintains frequency of the aberrant chromosome (Sabantsev et al. 1993).

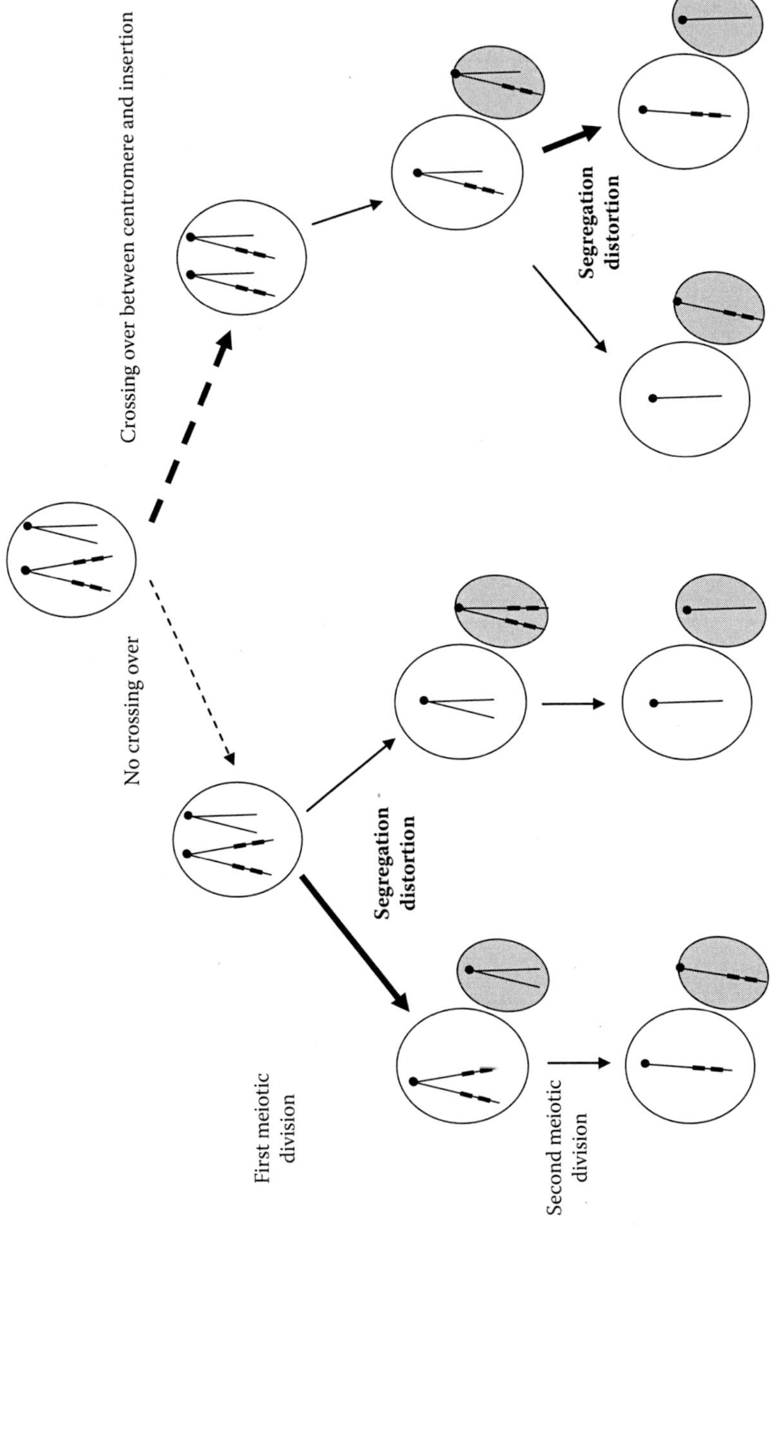
Crossing over between centromere and insertion
No crossing over
Segregation distortion
Segregation distortion
First meiotic division
Second meiotic division

There are other factors that may affect transmission ratios, including some deleterious mutations, maternal-fetal incompatibility in mammals, which usually do not cause considerable segregation distortions in populations. The human genome-wide study demonstrated that there is some "shift towards excess genetic sharing among siblings," with the average value of this shift for the autosomal loci in the order of 50.43% (Zölner et al. 2004). This data indicate that a few human genes show some degree of segregation distortion but the majority obey the rule.

Why is the first Mendelian law so common?

This question has been raised many times. Crow (1991) emphasized the importance of potentially numerous genetic elements which can suppress emerging genetic systems capable of causing stable and significant distortions of Mendelian ratios. He also paid attention to crossing-over which adds difficulties to establishing TRD or meiotic drive systems. However, genetic systems suppressing deviations from Mendelian segregations seem to be costly and in some cases might not be entirely effective. It was suggested that "meiosis is a delicate process seeded with general anti-drive devices" and this could explain the observation that in hybrids, which carry two very different sets of homologous chromosomes, there is little evidence of distortions but sterility is rather common (Hurst and Pomiankowski 1991). In other words if the meiotic process is overwhelmed by the differences between homologous chromosomes, which may cause Mendelian ratio distortions, it breaks down and causes sterility rather than leads to a segregation distortion.

Another possibility is related to interactions between an unlinked modifier locus and the locus in question which could experience segregation distortion. It was shown that natural selection disfavors the modifier alleles that promote segregation distortion and favors those alleles which have the opposite effect (Eshel 1985). Such selective pressures create evolutionary stability for equal segregation. Some linked loci may cause segregation distortion depending on their closeness to the locus in question, which affects the likelihood of recombination. Recent investigation of the problem concluded

Figure 5.5 (Opposite) Meiotic segregation in females heterozygous for chromosome 1 insertion. Chromosomes that carry the insertion are marked by two rectangulars. Oocytes are represented by large ovals and polar bodies by small and filled ovals. Two oocyte classes are formed during the first meiotic prophase after crossing-over: cells with recombinant chromatids, class I (right) which is most likely; and cells with nonrecombinant chromatids, class II (left). Segregation distortion takes place in the first meiotic division (left) and in the second division (right). The thick arrows indicate the more likely directions of chromosome segregation. Only oocytes contribute to the next generation. (Redrawn with minor modifications from Agulnik, Agulnik, and Ruvinsky 1990.)

that further studies are necessary for better understanding those selective forces that maintain Mendelian segregation (Úbeda and Haig 2005).

There may also be an exaggeration of exactness of equal transmission. To detect deviations from the first Mendelian law is not as easy a task as one may think, not only because such deviations are rare but also because this requires a large number of observations. The logic behind such a requirement is dictated by statistical laws. The error is smaller when the number of observations is larger. However, this statement is correct as far as individual events are random. Now presume that in a population there are two factors, one of which supports transmission of allele *A* and another allele *a*. In some individuals one of two alleles has better chances to be passed to the next generation, whereas in other individuals the alternative allele has transmission advantage. Adding the results of all observations in the population might lead to the expected 1:1 Mendelian ratio. Thus, heterogeneity of genotypes involved in certain crosses may mask deviations. As the observed deviations from equal transmission are usually small, in order of a few percentages, many of them could escape detection because opposite transmission distortions will negate each other.

It looks feasible that the number of minor segregation distortions in populations exceeds our current estimates but as soon as they shift frequencies in alternative directions, the total result is very close to Mendelian expectation. The principle of random mating, which is the basic assumption in population genetics, becomes an important factor stabilizing segregation ratios of interbreeding groups or populations.

The final conclusion we have to make is simple; rare deviations from Mendelian segregations may occur both in males (TRD) and females (meiotic drive). However, genetic systems with strong deviations from equal segregation are rare because in order to ignore the law such systems must overcome several considerable obstacles. In any case aggressive alleles or chromosomes have a limited opportunity to increase their frequencies in populations due to the disadvantages that they usually carry. It is possible that populations exhibiting significant segregation distortion may have lower fitness and if so this might be an additional explanation of their rarity.

Randomness rules

The experience accumulated by genetics clearly indicates that deviations from the first Mendelian law are rare and usually minor. It means that random transmission of alternative alleles is nearly always the case. This knowledge equips us well to forecast the ratio between two segregating alleles or chromosomes over a large number of offspring but does little to predict each individual case. For instance, if a couple wishes to predict the sex of a future child, the answer "either girl or boy" is hardly satisfactory. While the mechanisms involved in segregation of chromosomes look miraculously precise and even in a certain sense deterministic, the outcome is entirely random and universal. Forces of natural selection operating in populations maintain this status quo in all meiotically reproducing species. There is no need for a teleological explanation of such stability; rather equal transmission is the preferable state of genetic systems supported by natural selection. This stability is not static but dynamic and works well on the population level. Randomness of transmission is the basis for the Hardy–Weinberg principle (HWP), which is the cornerstone of population genetics processes (Tchetverikov 1926). Analogy with a thermodynamic description of complex and chaotic physical systems is quite appropriate here. Despite total uncertainty of an outcome of each single transmission event, our ability to calculate dynamics of genotypic and allelic frequencies in a population is high. This is the cause of the high predictive power of HWP.

Thus, when a single segregation/transmission event is considered, it usually has maximum uncertainty. However, numerous events particularly on the population level are highly predictable. Random segregation of homologous chromosomes in meiosis and randomness of matings and fertilization in populations represent an ubiquitous and important level of uncertainty. As soon as high regularity of segregation and transmission is violated by some TRD or meiotic drive systems, uncertainty is lessened; combinatorial variability is reduced; and fitness of such a population eventually could diminish.

As already mentioned, equal meiotic segregation together with crossing-over generate genetic variability on a gigantic scale. While mutations are the primary source of genetic variability and hence the fuel for Darwinian evolution, it is hypothetically conceivable that eukaryotic populations could evolve for a long time without new mutations using recombination, segregations, and random fertilization as sources of combinatorial variability. What really makes these combinatorial sources of genetic variability very much different is their biological origin. They were born together with meiosis, some 1.5 billion years ago and represent one of the most incredible advancements of biological evolution. Never before or after was a common biological process capable of generating randomness in unlimited quantities invented and widely used. Earlier we labelled such phenomena as organized randomness. Self-acceleration and self-diversification of eukaryotic life was the consequence of these evolutionary innovations. Regardless of the initial cause of evolution of meiosis, the result was quite unique—the development of a precise mechanism for equal segregation of homologous chromosomes capable of generating randomness and massive combinatorial variability.

The fact that combinatorial variability is universally used in the majority of eukaryotic species for a very long time provides an assurance of its usefulness and desirability. In other words eukaryotes should have sufficient gains from additional randomness which are not available for other forms of life. Would this not be true, one could expect that different means of reducing randomness were acquired during evolution. Among them we could anticipate transmission distortion, reduced or even prohibited recombination, and all sorts of assortative fertilization. Such deviations are known and we have already discussed systems that create significant transmission distortion. There are also examples of crossing-over suppression existing for a long time, including such extreme cases as total lack of recombination in *Drosophila melanogaster* males. These facts tell us that general rules operating in meiotically reproduced species can be ignored and cheaters may be prosperous. Still such systems are comparatively rare.

The existence of parthenogenetic species, which totally or partially abandoned the great achievements of eukaryotic evolution like meiosis, fertilization, and bisexuality, is an excellent demonstration that in certain conditions all this beautiful eukaryotic baggage can be lost and this does not necessarily lead to extinction at least for a lengthy evolutionary time. On the contrary in some taxons there are numerous parthenogenetic species that deserted the typical eukaryotic way of life and thrive despite this. Such species, however, usually occupy specific and narrow positions in ecosystems and rather represent side branches of the evolutionary tree, a kind of escape from the major road of progressive evolution. In taxons like birds and mammals, symbolizing remarkable morphophysiological advancement and adaptation to all major habitats, there are no examples of abandonment of meiotic reproduction. Certainly a straightforward conclusion that the combinatorial variability (and accompanied randomness) is the essential condition of such progress may not be warranted. But it probably can be said that the taxons which reached superior morphophysiological complexity used combinatorial variability extensively. In such species degree of randomness and uncertainty during the transfer of genetic information from generation to generation is very high indeed.

Even if the combinatorial variability was a collateral consequence of other emerging biological processes, this innovation was greatly used during eukaryotic evolution and had a dramatic impact on its speed and diversity. The same is applicable to randomness, which in this case has clear biological origin. This type of randomness is not known for other forms of life and certainly does not exist in nonorganic matter. Randomness resulted from meiotic segregation of homologous chromosomes and fertilization is usually close to the highest possible level and little could be done to change that. Perhaps this is a sign that uncertainty generated by such processes is welcome by eukaryotes.

Summary

Mendel's discovery of equal segregation and independent assortment created a foundation for genetics and explained the movements of genes and chromosomes from generation to generation. There was another side of this discovery. Mendel essentially found a highly organized source of randomness which exists in eukaryotic species that produce gametes meiotically. The regularity of segregation of homologous chromosomes during the first meiotic division is amazing taking into account possible allelic and structural differences between the homologous chromosomes. Each allele on average has ~50% probability to get into a gamete and uncertainty reaches maximum at this point.

There are rare situations when Mendelian segregation is significantly distorted. This includes transmission ratio distortion (TRD) and meiotic drive. In both cases an allele or a chromosome has a much higher chance to get into the next generation than an alternative allele or homologous chromosome. In extreme situations such non-Mendelian segregation dramatically reduces uncertainty because a particular allele or a chromosome might be delivered to the next generation with probability above 95%. Strong deviations from Mendelian segregation are usually controlled in populations by the balancing selection.

The equal transmission of alleles to the next generation is generally the rule and the question regarding the causes of its stability is pertinent. There are several not mutually exclusive explanations of such stability, including anti-TRD/meiotic drive devices, which act on meiotic as well as on population levels, and also influence crossing-over. It seems, however, that the current understanding of exceptional stability of equal segregation and transmission is not complete. The stability of the first Mendelian law has a dynamic nature and is supported by natural selection.

There are three types of randomness. Two of them, quantum randomness and molecular stochasticity, are well known for all forms of matter. The third one, randomness of allele transmission, is limited to eukaryotic species that use a meiotic process. Clearly this form of randomness is a product of evolution and originated about 1.5 billon years ago. This "biological" randomness is highly common for eukaryotes and is the exceptionally powerful source of combinatorial variability.

References

Agulnik, S.I., A.I. Agulnik, and A.O. Ruvinsky. 1990. Meiotic drive in female mice heterozygous for the HSR inserts on chromosome 1. *Genetics Research* 55:97–100.

Bauer, H., N. Véron, J. Willert, and B.G. Herrmann. 2007. The *t*-complex-encoded guanine nucleotide exchange factor *Fgd2* reveals that two opposing signaling pathways promote transmission ratio distortion in the mouse. *Genes & Development* 21:143–147.

Crow, J.F. 1991. Why is Mendelian segregation so exact? *BioEssays* 13:305–312.

Crow, E.F. and J.F. Crow. 2002. 100 years ago: Walter Sutton and the chromosome theory of heredity. *Genetics* 160:1–4.

Eshel, I. 1985. Evolutionary genetic stability of Mendelian segregation and the role of free recombination in the chromosomal system. *American Naturalist* 125:412–420.

Hartl, D.L., and D.J. Fairbanks. 2007. Mud sticks: On the alleged falsification of Mendel's data. *Genetics* 175:975–979.

Hurst, L.D., and A. Pomiankowski. 1991. Maintaining Mendelism: Might prevention be better than cure? *BioEssays* 13:489–490.

Johnson, L., C.S. Petty, and W.B. Neaves. 1980. A comparative study of daily sperm production and testicular composition in humans and rats. *Biology of Reproduction* 22:1233–1243.

Kusano, A., C. Staber, H.Y. Chan, and B. Ganetzky. 2003. Closing the (Ran)GAP on segregation distortion in *Drosophila*. *Bioessays* 25(2):108–115.

Lewontin, R.C. 1968. The effect of differential viability on the population dynamics of *t* alleles in the house mouse. *Evolution* 22:705–722.

Lyon, M.F. 2003. Transmission ratio distortion in mice. *Annual Review of Genetics* 37:393–408.

Mendel, G. 1865. http://www.esp.org/foundations/genetics/classical/gm-65.pdf p. 24.

Orel, V. 1996. *Gregor Mendel: The first geneticist*. Oxford, UK: Oxford University Press.

Pardo-Manuel de Villena, F., E. de la Casa-Esperon, T.L. Briscoe, and C. Sapienza. 2000. A genetic test to determine the origin of maternal transmission ratio distortion: Meiotic drive at the mouse *Om* locus. *Genetics* 154:333–342.

Ruvinsky, A. 1995. Meiotic drive in female mice: Comparative essay. *Mammalian Genome* 6:315–320.

Ruvinsky, A., A. Polyakov, A. Agulnik, H. Tichy, F. Figueroa, and J. Klein. 1991. Low diversity of t haplotypes in the eastern form of the house mouse, *Mus musculus L. Genetics* 127:161–168.

Sabantsev, I., O. Spitsin, S. Agulnik, and A. Ruvinsky. 1993. Population dynamics of aberrant chromosome 1 in mice. *Heredity* 70:481–489.

Sandler, L., and E. Novitski. 1957. Meiotic drive as an evolutionary force. *American Naturalist* 41:105–110.

Schimenti, J. 2000. Segregation distortion of mouse t haplotypes: The molecular basis emerges. *Trends in Genetics* 16:240–243.

Schneider, T.D. 2005. Information theory primer. http://www.cbs.dtu.dk/dtucourse/27611spring2006/Lecture04/Information_Theory_Primer.pdf. (accessed April 18, 2008).

Seitz, A.W., and D. Bennett.1985. Transmission distortion of t-haplotypes is due to interactions between meiotic partners. *Nature* 313:143–144.

Tchetverikov, S.S. 1926. O nekotorykh momentakh evoliutsionnogo protsessa s tochki zrenia sovremennoi genetiki. *Zhurnal eksperimental'noi biologii*, ser. A, 2(1):3–54 (in Russian). On certain aspects of the evolutionary process from the standpoint of modern genetics. ([1961].Transl. of 1926 paper.) *Proceedings of the American Philosophical Society* 105:167–195.

Úbeda, F., and D. Haig. 2005. On the evolutionary stability of Mendelian segregation. *Genetics* 170:1345–1357.

Weiling, F. 1986. What about R.A. Fisher's statement of the "too good" data of J.G. Mendel's Pisum paper? *Journal of Heredity* 77:281–283.

Zöllner, S., X.Wen, N.A. Hanchard, M.A. Herbert, C. Ober, and J.K. Pritchard. 2004. Evidence for extensive transmission distortion in the human genome. *American Journal of Human Genetics* 74:62–72.

chapter six

Random genetic drift and "deterministic" selection

> The success or failure of a mutant gene in a population is dependent not only on selection but also on chance.
>
> **Motoo Kimura**
> *On the Probability of Fixation of Mutant Genes in a Population (1962)*

The discovery of genetic drift

Classical Darwinian views, as is well known, are based on two pillars: random hereditary variability (Chapter 2) and natural selection, which traditionally was considered the driving force leading to adaptations and, hence, a nonrandom factor. At the time of the rediscovery of Mendel's laws and the establishment of genetics, little was known about hereditary variability and little direct evidence of natural selection was available. Not surprisingly many early geneticists were initially quite skeptical about Darwin's hypothesis and believed that genetics alone would be sufficient to explain evolution. Steadily the situation began to change, and the first sign of this change was the emergence of the Hardy–Weinberg Principle (HWP) which became a solid foundation for understanding population processes (Hardy 1908; Weinberg 1908). However, the gestation period took some time and population genetics, as a well-defined discipline, appeared only in the late 1920s (see Box 6.1).

According to HWP, in a large population the frequencies of alleles in the previous generation determine the frequencies of genotypes in the following generation assuming random matings. HWP also states that without external interventions frequencies of alleles and genotypes remain unchanged in successive generations. Despite the obvious importance, HWP was not frequently applied until Sergey Tchetverikov (1926) in Russia explicitly used it for explaining genetic processes in natural populations of *Drosophila*. It was the starting point for experimental population genetics (Adams 1970). From the early 1930s this new field of genetics began to flourish in the United States in the work of Theodosius Dobzhansky, his colleagues and students.

BOX 6.1 THE HARDY–WEINBERG PRINCIPLE

A discussion about allele dynamics in populations had begun soon after Mendelian laws were rediscovered in 1900. One of the arguments was that frequencies of dominant alleles should increase in populations. Then the American geneticists William E. Castle showed that without selection, the genotype frequencies would remain stable. Nevertheless Reginald Punnett thought that a formal solution of the problem is desirable and discussed the problem with British mathematician Geoffrey H. Hardy, who resolved the problem and on July 10, 1908 published a letter in section "Discussion and Correspondence" (28:49–50) of the American journal *Science*. A fragment of this letter follows:

Suppose that Aa is a pair of Mendelian characters, A being dominant, and that in any given generation the number of pure dominants (AA), heterozygotes (Aa), and pure recessives (aa) are as p:2q:r. Finally, suppose that the numbers are fairly large, so that mating may be regarded as random, that the sexes are evenly distributed among the three varieties, and that all are equally fertile. A little mathematics of the multiplication-table type is enough to show that in the next generation the numbers will be as $(p+q)^2{:}2(p+q)(q+r){:}(q+r)^2$*, or as* $p_1{:}2q_1{:}r_1$*, say.*

The interesting question is — in what circumstances will this distribution be the same as that in the generation before? It is easy to see that the condition for this is $q^2 = pr$*. And since* $q_1^{\,2} = p_1r_1$*, whatever the values of p, q, and r may be, the distribution will in any case continue unchanged after the second generation.*

For the next 3-4 decades this principle was known in the English speaking countries as *Hardy's law*. It was only in 1943 when famous geneticist Curt Stern, who fled Germany, raised the point that the same principle had first been formulated independently and also in the same time by the German physician Wilhelm Weinberg. There were several reasonable attempts to associate Castle's name with the principle in order to recognise his contribution, but unfortunately this was too late, Hardy-Weinberg principle became a common name.

Sources: Crow J.F. 1999. Hardy, Weinberg and Language Impediments. *Genetics*, 152:821–825; Edwards A.W.F. 2008. G. H. Hardy (1908) and Hardy–Weinberg Equilibrium. *Genetics*, 179: 1143–1150.

Meanwhile, starting from the 1920s, the foundations of theoretical population genetics were built independently mainly by the efforts of three great scientists. Ronald Fisher, the outstanding British statistician and geneticist, published in 1930 a highly influential book *The Genetical Theory of Natural Selection*. Fisher's fundamental theorem of natural selection was among the significant theoretical advances. John Haldane was another outstanding British scientist whose contribution to emerging population genetics was diverse and long lasting. In the book *The Causes of Evolution*, published in 1932, he summarized the results of his investigations in the mathematical theory of population genetics. The third researcher, American geneticist Sewall Wright, was not only the major contributor to population genetics but a towering figure of twentieth century genetics. His publications continued for 76 years from 1912 to 1988 (Crow 1988; Figure 6.1, Box 6.2).

Figure 6.1 Sewall Wright (1889–1988), outstanding American geneticist, one of the creators of theoretical population genetics. The theory of genetic drift is his major achievement. He was born in Melrose, Massachusetts, and died in Madison, Wisconsin. (Courtesy of the Special Collections Research Center, University of Chicago Library.)

BOX 6.2 SEWALL G. WRIGHT

Sewall Green Wright was born in 1889 in Melrose, Massachusetts. Three years later his family moved to Galesburg, Illinois, when his father got a teaching job at Lombard College. From childhood Sewall Wright had a great interest in mathematics and biology. After graduation from high school in 1906 he enrolled at Lombard College to learn mathematics and then he studied at the University of Illinois at Urbana. In 1912 Wright graduated with a master's degree and moved to Harvard to continue his education with Professor William Castle, a pioneer in genetics research. At Harvard Wright conducted well-known studies on the genetics of coat color in mammals, working with guinea pigs. He earned a doctorate in zoology (Sc.D.) in 1915. Then young but very well prepared, Wright took a position at the U.S. Department of Agriculture, where he worked for the next ten years. His major work, devoted mainly to investigation of inbreeding, mating systems, path coefficients, and physiological genetics, was carried out mainly in those years.

In 1925 Wright joined the department of zoology at the University of Chicago, where he worked for nearly thirty years. His major achievement during this period was the development of the theory of genetic drift. Wright believed that the interaction of genetic drift and other evolutionary forces was an important component of the evolutionary process. Another brilliant achievement was the development of the shifting balance theory. By the end of his career in Chicago theoretical population genetics had become a well-recognized and well-developed discipline and Wright's input to this spectacular progress was massive. After retirement at the age of 65 he moved to the University of Wisconsin (1955), where he remained active as an emeritus professor for a third of a century until his death in 1988 in Madison. A mathematic model of evolutionary genetic processes which is well integrated with the ideas of Darwin and Mendel is an outstanding contribution of Sewall Wright to modern biology.

Sources: Crow J.F. 1988. Sewall Wright (1889–1988). *Genetics* 119:1–4; Provine, W.B. 1989. *Sewall Wright and evolutionary biology.* Chicago: University of Chicago Press.

HWP, which became a point of departure for population genetics considerations, is often compared with the first Newton law as it too describes inertia. For a large random mating population it means that the frequency of alleles and genotypes remains unchanged from generation to generation in the absence of external forces. HWP does not specify the size of such an ideal population as it is considered practically infinite. Random matings or, in other words, random fertilization events, is the critical condition for Hardy–Weinberg equilibrium. In many real populations HWP works well but at rather short intervals. Statistical deviations became more sizable when numerous consecutive generations are considered. In small and particularly very small populations deviations from HW expectations are high and frequencies of alleles and genotypes drift from generation to generation usually in an entirely random fashion.

The concept of genetic drift is among Wright's finest achievements. His paper "Evolution in Mendelian Populations" (1931) laid the foundation for a novel view of random processes in populations. During the 1920s Wright realized that in small populations sampling errors in basic biological processes responsible for allele transmission as well as formation of genotypes is significant. Accordingly the frequency of an allele or a genotype in a small population can rise or fall considerably and rapidly (Figure 6.2 a-d). This purely random process is dependent on a number of reproducing individuals in a population and time, which is usually measured in generations. In small populations the intensity of the drift is high. An allele can be fixed in a population, when its frequency reaches a value of 1; it also means that an alternative allele is lost. These two points are critical because further genetic drift of allelic frequencies in the population becomes impossible until a new allele appears in the population due to a mutation or migration. Wright's conclusion was deemed to be very important as it discovered a previously unappreciated factor of evolution and this was a fundamental contribution to population genetics.

At the same time and independently, similar views were developed by Russian geneticists D.D. Romashov and N.P. Dubinin in 1931 to 1932 (Dubinin 1931; Dubinin and Romashov 1932), which were inspired by population genetics studies initiated by Tchetverikov (Adams 1970). Andrei N. Kolmogorov, one of the leading mathematicians of the twentieth century, who among other achievements made a solid contribution to the probability theory, was well aware of their research as he was Romashov's classmate and knew Dubinin from his college years (Tihomirov 2003; Adams 1970). At about the same time Kolmogorov developed a mathematical theory of stochastic changes, which is relevant to population genetic processes. The time-dependent solution of the forward Kolmogorov (Fokker–Planck) equation describing the stochastic changes in allelic frequencies in a finite population, developed by Kimura (1955) a quarter of century later, became an important advancement of the genetic drift theory (Ewens and Heyde 1995).

Much earlier both Fisher (1922) and Haldane (1927) had published their theoretical results showing that in finite size populations stochastic effects are inevitable. The question was how to treat this randomness. The following citation provides a clear account of the differences in the views held by Fisher and Wright:

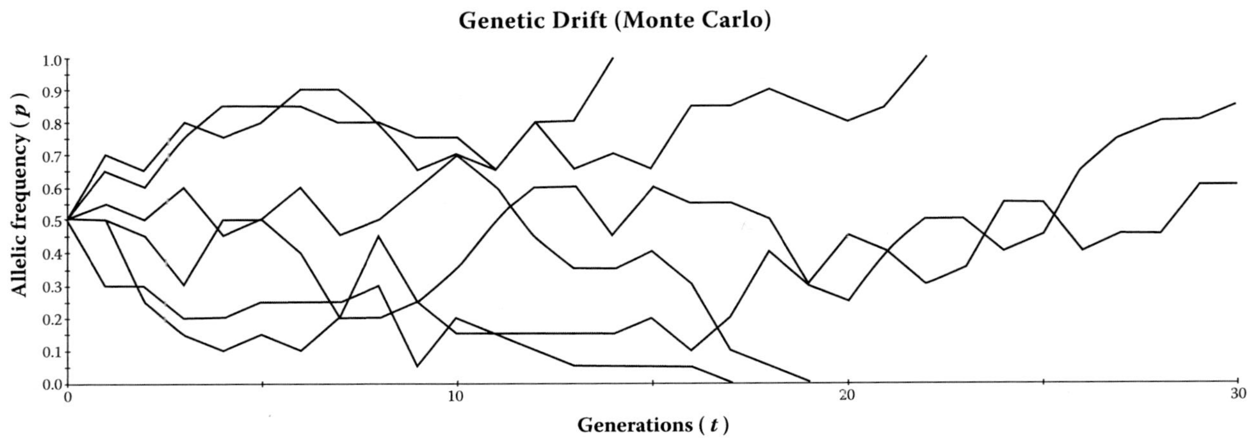

Figure 6.2 Monte Carlo computer simulations of allelic frequencies in Mendelian populations using POPULUS 5.4 program (Alstad 2007) for six simultaneous independent trials. Fitness of all alleles is the same. A Mendelian population can be defined as an imaginary group of interbreeding, bisexually reproducing individuals. Random matings are considered to be very typical for Mendelian populations. Each simulation is completely random and the graphs represent the outcomes of a few simulations. (a) Starting allelic frequency, $p = 0.5$; population size, $N = 10$. (b) Starting allelic frequency, $p = 0.5$; population size, $N = 100$. (c) Starting allelic frequency, $p = 0.05$; population size, $N = 10$. (d) Starting allelic frequency, $p = 0.05$; population size, $N = 100$.

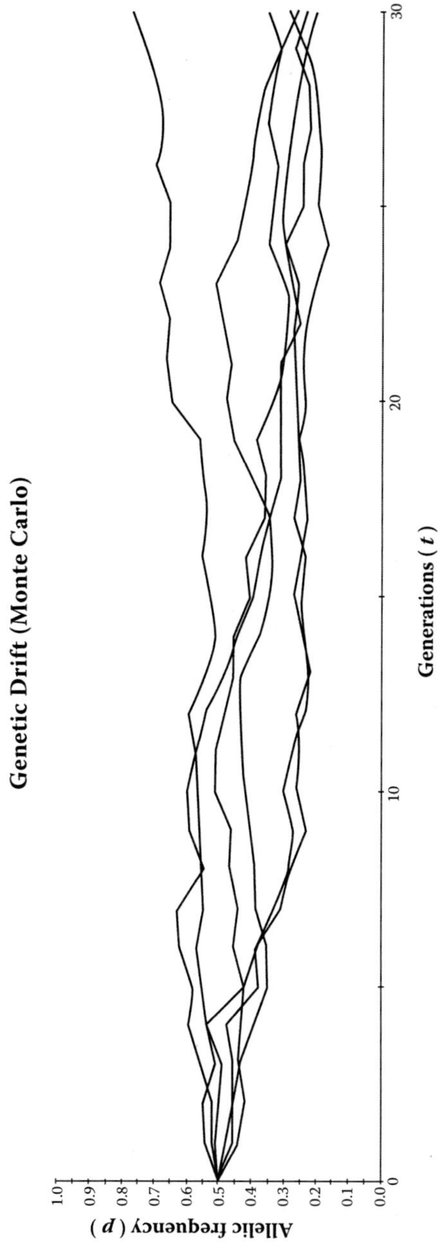

Figure 6.2b (Continued).

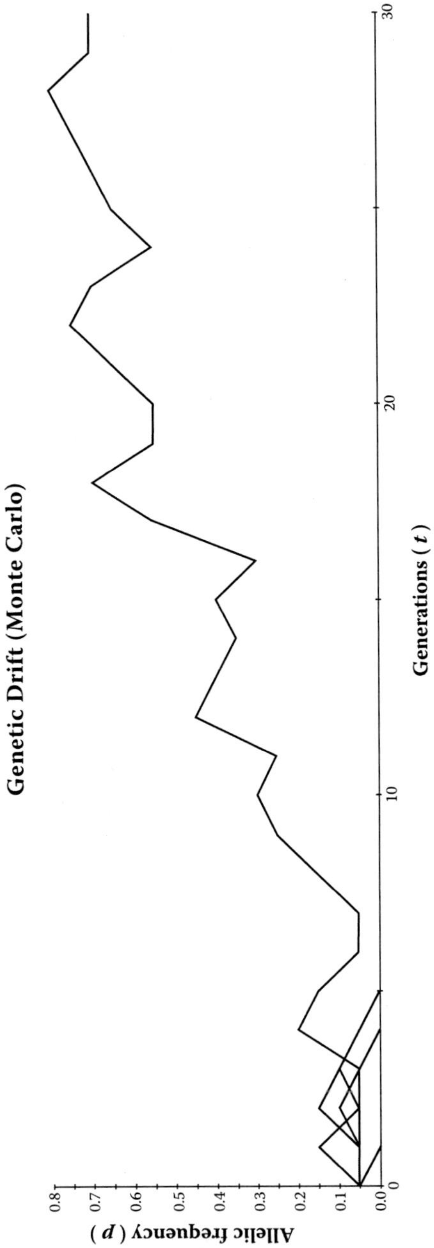

Figure 6.2c (Continued).

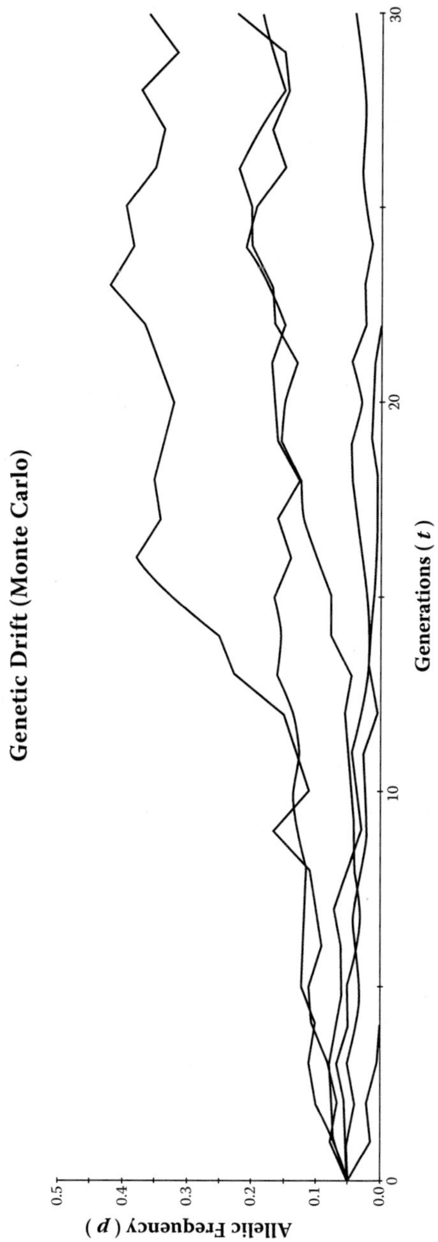

Figure 6.2d (Continued).

> The views of Fisher and Wright contrast strongly on the evolutionary significance of random changes in the population. Whereas, to Fisher random change is essentially noise in the system that renders the determining processes somewhat less efficient than they would otherwise be, Wright thinks of such random fluctuations as one aspect whereby evolutionary novelty can come about by permitting novel gene combinations (Crow and Kumura 1970).

The essence of the debate that followed was about the role of stochastic processes in real populations, frequency of occurrence of small populations, rates of allele fixation, and possible interactions of genetic drift and selection. This discussion continued with varying intensity well into the 1980s and possibly could reignite again. The theory of genetic drift developed by Sewall Wright was a very important attempt to incorporate stochastic processes in theoretical population genetics and evolutionary biology.

Random events, like tossing a coin, do not depend on previous outcomes or affect the following events. Each individual trial is entirely independent from all others. The same is correct for individual random genetic events. However, if we consider a population, the conclusion is different. Assume that in a small population with two alleles there were several random events that increased the frequency of allele A_1 and correspondingly decreased the frequency of alternative allele A_2. This chain of random events certainly does not affect the probability of individual outcomes in subsequent genetic trials but it may increase the likelihood of allele A_1 being lost and allele A_2 becoming fixed in the population (or vice versa). This means that past random genetic evens can affect the future genetic structure of a population.

In small populations this scenario is particularly pertinent and fixation of some alleles and loss of others may occur during a relatively short period of time (Figure 6.2a). In larger populations the intensity of genetic drift declines as shown in Figure 6.2b; in this particular computer simulation neither fixation nor loss of an allele occurred over one hundred generations, but there are a few cases where fixation or loss of an allele is more likely. The outcome of genetic drift experiments also depends on the initial frequency of an allele in question. If the starting frequency of an allele is low (or high) there is a much higher likelihood of its loss (or fixation). Figure 6.2c shows that if in small populations the starting frequency of an allele is 0.05, most such alleles are lost very quickly. If the population is larger, the same trend can be observed but it takes much longer to lose or fix alleles. This means that the frequencies of alleles in small populations change more dramatically in just a few generations. Even alleles with lower fitness occasionally may have higher frequencies than their counterparts. In other words frequencies of alleles in small populations are not highly correlated with their adaptive values.

The mathematical approach introduced by Wright was successfully put to the test when he and Dobzhansky joined efforts in describing genetic drift in natural populations (Dobzhansky and Wright 1941). The role of isolation and small populations became a prominent topic in population genetics and evolutionary studies in the decades that followed. Ernst Mayr, one of the major architects of the modern evolutionary synthesis, embraced the idea of genetic drift and it became an essential part of his theory of allopatric speciation (Mayr 1963). Small breeding populations or demes, using Wright's terminology, seem to be quite common in many species. In case of their temporary isolation losses of some alleles due to random events are widespread, which reduces heterozygosity and increases homozygosity.

Two similar situations may amplify changes occurring in small populations. One of them is the so-called bottleneck effect, when population size shrinks significantly for some time. Another is the founder effect, when a small population becomes an ancestor to a large population or a group of populations. A typical consequence for each of these situations is a considerable change of allelic and genotypic structure of the populations due to entirely random events. Natural selection can modify the outcomes depending on fitness of alleles or genotypes but nonetheless genetic drift is a random factor capable of changing the frequencies of alleles regardless of their fitness. As expected, genetic drift may also reduce the efficiency of natural selection. Theoretical analysis estimates that "the fixation rate of a favourable mutant is significantly lower when genetic drift is considered; fixation probability is reduced by over 25% for realistic" experiments (Heffernan and Wahl 2002).

In extreme situations, when the frequency of an allele is very high or very low, selection is not effective. For instance, when the frequency of recessive allele *a* is low [for instance $q(a) = 0.01$] and only homozygotes *aa* express a trait different from wild type, selection in favor of this trait is inefficient because the homozygotes occur very rarely in a population [$q^2\,(aa) = 0.0001$]. This also means that allele *a* is "hidden" in heterozygotes *Aa*, which have wild phenotype and thus escape selection. Similar circumstances occur when a dominant allele has high fitness. Selection becomes ineffective as soon as the frequency of this allele is getting high [for instance $p(A) = 0.99$] because a chance for homozygotization of rare recessive allele *a* is really low [$q^2(aa) = 0.0001$] and the rare recessive allele "hides" from selection behind the dominant allele in heterozygotes. In both these cases genetic drift can lead to a quick loss or fixation of an allele in small populations in just a few generations and this is a result of totally random sample variation, which has nothing to do with fitness of an allele.

Many natural populations are large enough to reduce the influence of random sampling errors and, hence, genetic drift, quite significantly. Still all biological populations have finite population sizes and this means that genetic drift always exists and, given a lengthy time, will change the frequencies of alleles in populations. This is very much different from natural selection, which links changes in the genetic structure of a population with adaptive and reproductive values of individuals. Genetic drift and natural selection being distinct factors of evolution affect frequencies of alleles in the same populations and because of this do not act in isolation. The degree to which alleles and genotypes are influenced by genetic drift and selection depends on population size; in smaller populations contribution of genetic drift is much more pronounced.

Neutral mutations in evolution

In 1954 Motoo Kimura joined James Crow's laboratory at the University of Wisconsin; Sewall Wright moved to Wisconsin at the same time. These circumstances had an exceptionally positive impact on further development of the mathematical theory of genetic drift (Box 6.3, Figure 6.3). Over the next ten to twelve years there was great progress in mathematical description of stochastic processes in finite populations. This included finding the solution to several key equations of diffusion theory, which allowed estimations of allele fixation probability in more complex situations, such as multi-allele loci, dominance, and selection. Among other tools, Kimura used the Kolmogorov backward equation, which considerably facilitated studies of allele losses or fixations. He also estimated an average time required for allele fixations or losses. The revival of classical mathematical population genetics definitely took place. One of the important and ground-shaking qualitative conclusions that followed from these

BOX 6.3 MOTOO KIMURA

Motoo Kimura was born in 1924 in Okazaki, Japan. From childhood he was very excited by living organisms, particularly plants. This interest led him in 1944 to Kyoto Imperial University, where he studied botany and became involved in plant cytogenetics. After graduation he worked as assistant to the famous cytogeneticist Professor Hitoshi Kihara. In the following years his interests steadily shifted toward population genetics and he learned the classical works of Fisher, Haldane, and Wright, as well as relevant mathematics. His cousin-in-law, Professor Matsui Tamura, was a mathematical physicist at the same university and supported Kimura's mathematical interests. In 1949 Kimura obtained a research position in the newly established National Institute of Genetics in Mishima, with the help of Kihara. He worked at the institute for the next forty years until his retirement. In 1953 Kimura received a fellowship to study in the United States, which eventually led him to the University of Wisconsin, Madison, where he began studies under the supervision of James Crow. Sewall Wright moved to Wisconsin in the same year, which provided a unique opportunity for the young researcher. Kimura wrote later that the time in Madison was one of the most productive periods in his scientific career. He finished his Ph.D. in 1956 and returned to Japan. Kimura published numerous papers in key journals and rapidly became a world leader in mathematical population genetics, being able to solve several problems unfinished by Fisher and Wright. Among them was finding the complete time-dependent solution of the forward Kolmogorov (Fokker–Planck) equation describing stochastic changes in finite populations. In 1961 Kimura came back to Crow's laboratory at Wisconsin and spent two years there. During this time he wrote several influential papers. Up until the mid-1960s Kimura was a strong neo-Darwinist. However, his purely mathematical work on allele fixation probability in populations steadily developed the theoretical background for the neutral theory of molecular evolution. Despite the fact that Kimura continued his rigorous and intensive theoretical population research for the next two decades, many believe that his major achievement after 1968 was the development and advocacy of this new theory. The exponentially growing amount of molecular data made his quest even more potent. The controversy that surrounded the idea of neutral evolution steadily subsided and a new vision of the evolutionary process emerged, which firmly included neutral evolution on a molecular level. Kimura was particularly touched to be awarded the Darwin Medal by the Royal Society (UK), as he never intended to diminish the idea of natural selection. Motoo Kimura died on November 13, 1994, on his seventieth birthday, in Mishima, Japan.

efforts was the realization that an average rate of allelic substitutions for selectively neutral mutations is equal to a mutation rate at the locus.

At about the same time neo-Darwinian views reached its highest point. There was a widespread opinion that fixation of mutations and major changes of allelic frequencies are usually caused by natural selection. An alternative view, that genetic drift is the major factor leading to fixation

Figure 6.3 Motoo Kimura (1924–1994), outstanding Japanese geneticist. Kimura made major contributions to stochastic theory and became one of the creators of the neutral theory of molecular evolution. He was born in Okazaki and died in Mishima, Japan. (Courtesy of *Genetics*, the journal of the Genetics Society of America.)

of alleles, was not common. The results obtained by Kimura and other talented researchers supported this alternative point of view. However, a lack of sufficient experimental data was an impediment. Meanwhile due to Kimura's efforts a huge gap between the substitution estimates based on cost of natural selection and genetic drift became more apparent. A controversy over the causes of polymorphism in natural populations sharpened. A comprehensive review of selectionism and neutralism in molecular evolution written by M. Nei (2005) describes this complex problem. Steadily, beginning in the late 1950s, molecular data began to emerge. Information concerning rates and types of amino acid substitutions and then nucleotide substitutions became more available. This data brought much needed experimental support to the genetic drift concept.

Eventually Kimura (1968) and then King and Jukes (1969) put forward ideas of the neutral theory of molecular evolution. Leaving aside the claim about non-Darwinian evolution made by King and Jukes, the main points of this theory were as follows. Neutral or nearly neutral mutations are the most common type of mutations existing in natural populations on a molecular level. If so, the great majority of polymorphisms in populations have a transient nature, which means that alleles are on their way to either fixation or loss. These processes are governed by random genetic drift. Cases of balanced polymorphism, when frequencies of alleles in populations are regulated by natural selection, are rather rare. This theory was applied to molecular evolution and did not deny the role of natural selection in evolution of morphological and physiological traits.

There were plenty of debates concerning the definition of a neutral mutation. According to the population genetic definition proposed by Kimura, if selection coefficient s is less than or equal to $1/2N_e$, where N_e is the effective population size or simply a number of reproductively capable individuals in a population, the allele or mutation is considered to be neutral. There are critical comments regarding such a definition (Nei 2005). Skipping this discussion it can be said that a commonly accepted definition does not exist but there is an estimate satisfactory for many researchers. If a mutation does not change the function of a gene or protein appreciably and the selection coefficient for such a mutation does not exceed 0.001, the mutation can be categorized as neutral or nearly neutral. There is sufficient confidence that mutations or alleles in this category are fixed or lost in populations due to random genetic drift rather than selection.

The rate of molecular evolution of certain proteins or genes, while it is not perfectly constant, demonstrates a significant degree of stability over lengthy evolutionary periods. This phenomenon is called the molecular clock, and it was introduced by Emil Zuckerkandl and Linus Pauling in their visionary paper in 1962. As Kimura and Ohta (1971) wrote:

> Probably the strongest support for the theory (*the theory of molecular evolution, AR*) is the remarkable uniformity for each protein molecule in the rate of mutant substitutions in the course of evolution.

While Kimura carefully tried to separate mechanisms of molecular evolution from phenotypic evolution (Takahata 2007), Nei put forward logical and factual arguments in favor of a dominant role of mutations in evolution. In his most recent publication he concluded:

> Novel mutations may be incorporated into genome by natural selection (elimination of pre-existing genotypes) or by random process such as genetic or genomic drift. However, once the mutations are incorporated in genome they may generate developmental constrains that will affect the future direction of phenotypic evolution. It appears that the driving force of phenotypic evolution is mutation, and natural selection is of secondary importance. (Nei 2008)

It is unlikely the majority of biologists are ready to embrace the view that natural selection is of secondary importance. Nevertheless, it is clear that random genetic drift is a powerful evolutionary force responsible for numerous fixations and losses of alleles. It also looks plausible that some of these events may influence future pathways of evolutionary processes.

The question of constraints, whether on a molecular level or beyond, and their role in evolution still needs further investigation. It has been known since the early days of molecular genetics that variation in genome size between species is enormous. This is true not only for the large taxons like plants, which have >1000-fold variation in the genome size, but also for groups of closely related species (Knight, Molinari, and Petrov 2005). The great bulk of this variable DNA content consists of different types of nongenic repeats and unique sequences. On the contrary the number of genes changes very little between species, only about fivefold between humans and yeast, distinct representatives of the eukaryotes. A number of facts and observations indicate that there is a constant genome minimization selective pressure. Species with highly inflated genomes usually end up in narrow ecological tunnels with a greatly diminished opportunity for future evolutionary success. This is a good illustration of how genomic features likely accumulated due to either random drift or local selective advantages, and despite mild long-term negative selection pressures, may create tight constraints for future evolutionary pathways.

Is natural selection deterministic?

The molecular data that became so abundant during the last two to three decades provided an excellent opportunity to study natural selection. Among many other things it was found that the nonsynonymous substitutions, which change the sense of codons, are much less frequent in those sections of genes that code for functionally critical parts of proteins, than synonymous substitutions. According to Nei (2008) in highly conserved sections of *HOX* genes 99.7% of nonsynonymous mutations are eliminated by selection, while in genes coding for more typical proteins only about 85% of nonsynonymous mutations are purged. Regardless of whether natural selection is considered as "an editor or a composer" there is an increasing flow of data strongly supporting an active role of natural selection on the molecular as well as on the phenotypic levels.

Natural selection is a complex population process that effectively increases the frequency of individuals with favorable heritable traits and hence makes individuals with less favorable traits rarer. In other words a cause for a particular selective pressure always exists. Does this mean that natural selection is strictly a deterministic factor? Probably not.

Timofeeff-Ressovsky used to joke that natural selection does not guarantee immortality. There are several well-known reasons why the correlation between the general direction of natural selection and each real outcome is not high all the time. Earlier in Chapter 4 we discussed the connection between genotype and phenotype and concluded that repeating exactly the same developmental trajectory for a genotype is practically impossible. As a result, each genotype could potentially produce a multitude of more or less similar phenotypes. Natural selection always acts on the phenotypic level and its efficiency depends on the correlation between the chosen phenotypes and the genotypes behind them. In other words developmental noise, which is an unavoidable and random factor, significantly restricts deterministic aspects of natural selection. Another relevant and essential point is the opportunistic nature of selection. Constant and countless changes in environmental conditions affect the direction and intensity of natural selection. What was highly desirable yesterday might not be very important tomorrow or next year. These often unpredictable changes inevitably affect the process of natural selection and cause uncertainty of selection outcomes.

Another factor influencing the process of natural selection is pure chance. First of all, some best genotypes might be unable to reproduce or survive due to a purely random combination of circumstances. Selective advantage is not a guarantee for the individual with favorable traits, it is just a higher probability to survive and reproduce. Second, fixation of a mutation due to random genetic drift may create potential constraints, which could influence the future direction and intensity of selection. This question is discussed briefly in the following section. And it is also essential to realize that the future requirements of natural selection cannot be known in advance because the preceding random events have not happened as yet. Both direction and intensity of future natural selection might be shifted dramatically by a few random events and the whole course of evolution for a particular species or entire biosphere might change. This could be as small as random fixation of a mutation in a species or as big as the impact of a huge meteorite. Constant fluctuation in direction and intensity is a very typical feature of natural selection.

The probabilistic nature of evolution, which is the basic element of the Darwinian paradigm, results not only from random mutations and random genetic drift. The uncertainty of natural selection is also the major contributor to unpredictability of evolutionary processes. Thus, despite the central role of natural selection in navigating evolutionary processes through a contingency of constantly changing conditions, natural selection itself is not a purely deterministic factor. Although this conclusion is nearly as old as the Darwinian approach itself, one should not forget it in order to avoid a simple deterministic description of natural selection.

Natural selection indeed is a nonrandom force but it is not strictly deterministic either.

Adaptations and stochastic processes in evolution

The basic tenet of evolutionary biology is recognition of the key role of natural selection in developing adaptations. The definitions of adaptation vary. For instance, Wikipedia gives the following definition: "An adaptation is a positive characteristic of an organism that has been favored by natural selection and increases the fitness of its possessor" (http://en.wikipedia.org/wiki/Adaptation). Here the connection between adaptation and natural selection is straightforward; it effectively excludes a possibility that some adaptations might not be the result of natural section. This is a very common way to define an adaptation. Other definitions are less restrictive: "The adjustment or changes in behavior, physiology, and structure of an organism to become more suited to an environment" (http://www.biology-online.org/dictionary/Adaptation) or "In the evolutionary sense, some heritable feature of an individual's phenotype that improves its chances of survival and reproduction in the existing environment" (http://www.everythingbio.com/glos/definition.php?word=adaptation). A majority of researchers probably would endorse those definitions that link adaptation and natural selection.

Since the late nineteenth century there have been many successful attempts to accumulate evidence of natural selection. Genomic studies provided an outstanding opportunity to check this idea once again and the results strongly support an active role of natural selection in adaptogenesis (Williamson et al. 2007). Below are a couple of examples demonstrating the connection between selection and adaptation on a molecular level. The first one is based on analysis of a genomic dataset of 1.2 million single-nucleotide polymorphisms genotyped in three human racial groups. This study identified 101 regions with very strong evidence of a recent selective sweep, which is the reduction in nucleotide variation in DNA surrounding some mutations. Numerous genes, which were likely involved in the adaptation process, have been identified within these regions. While the average length of the regions was modest (~100 kb), linkage dramatically increased the sphere of their influence to as much as 10% of the human genome. The recent origin of this selective sweep and the size of the genomic regions involved confirm the Darwinian connection between adaptive changes and natural selection. Another study, which compared *Drosophila simulans* and *D. yakuba*, showed that ~45% of all amino acid substitutions have been fixed by natural selection (Smith and Eyre-Walker 2002). Thus, adaptive substitutions are common in these species.

The question, however, is not whether natural selection causes adaptive changes. This is a generally accepted point of view based on numerous facts. The question is whether all adaptive changes are caused by selection. Surely it seems unreasonable to expect that random factors, like genetic drift, could promote adaptations. This matter has been discussed in several publications lately. A theory was suggested that quantitatively describes how neutral evolution leads to marginally stable proteins and it provides a mathematical explanation of how protein biophysics influences evolution (Bloom, Raval, and Wilke 2007). Proteins have to satisfy two basic requirements, the ability to perform certain biological function (for instance to act as enzymes) and the ability to fold in certain two- and three-dimensional structures. The majority of proteins are marginally stable, that is, they have a relatively narrow range of free energy in which they fold in correct structures. Slow accumulation of mutations during the evolution of many homologous proteins diverge their amino acid sequences significantly; still they show a surprisingly high degree of conservation of both three-dimensional structures and functions. Such an outcome is expected in the case of neutral evolution, when fixed mutations usually have a very small impact. Both experimental and theoretical evidence support this point of view. However, there is a possibility for some neutral mutations to obtain adaptive significance. A perfectly neutral mutation, which does not measurably affect either structure or function of a protein, may influence that protein's tolerance to subsequent mutations. For instance, it may become incompatible with a range of potential mutations and, if so, limits an opportunity for further mutational changes in a part or entire protein. It means that a mutation at the time of fixation could be perfectly neutral and does not affect fitness in any measurable way; still it might or might not influence the future direction of evolution.

Proteins do not evolve in isolation; instead they are included in complex networks. It was found that the most stable proteins usually occupy central positions (hubs) of such networks. A concept of neutral networks was developed, which allowed quantifying the extent of mutational robustness or, in other words, "the insensitivity of the phenotype to mutations" (van Nimwegen, Crutchfield, and Huynen 1999). Another interesting result obtained within the neutral networks theory is the connection between protein stability and mutational robustness (Bloom, Raval, and Wilke 2007). Traditionally homeostasis of phenotypes is considered a result of stabilizing selection, which through a number of mechanisms is able to increase robustness of phenotypes to external and internal factors. The idea of neutrally evolved mutational stability in some proteins certainly does not dismiss the classical views; it rather suggests an additional explanation which assumes that neutral evolution may influence adaptively important traits like mutational robustness and protein stability.

Another related problem is the origin of genetic networks. Lynch (2007) questioned the common view that "the global features of genetic networks are moulded by natural selection" and concluded that no formal demonstration of the adaptive origin of any genetic network has been available so far. His analysis demonstrates that numerous features of known transcriptional networks could be produced by genetic drift, mutation, and recombination. According to Lynch three observations motivate the hypothesis that regulatory-pathway evolution on many occasions is driven by nonadaptive processes. First of all, he indicates the "unnecessary" complexity of genetic pathways, which are metaphorically labelled as "baroque" structures. Indeed, the analysis of many pathways reveals the presence of steps or elements that are conserved remnants of the past rather than the essential components of modern systems. It is unclear whether the long pathways consisting of genes activating or inhibiting the following elements have any advantage over more simple pathways. A removal or silencing of a pair of alternating elements may not have any physiological effect. Second, shifts in regulatory infrastructure, like transcription factor binding sites, often occur without noticeable changes in phenotypes. And finally, as Lynch shows, "the ratio of rates at which regulatory sites are lost and gained" by fixation of mutations is critically important for determining the evolutionary patterns of genetic pathways. All this raises the question "whether natural selection is necessary or even sufficient for the origin of many aspects of gene-network topologies" (Lynch 2007). A clear answer to this and similar questions is important for better understanding of the roles that randomness, natural selection, and self-organization play in the origin and evolution of life.

As Brian Kinghorn put it, genetic networks have a robust design but they are not designed. If such genetic networks indeed were engineered, simpler constructions could be implemented without losing their essential qualities. The redundant complexity of genetic networks reflects convoluted and lengthy evolutionary pathways. Since the earliest days of genetics research it was known that an acceptable phenotypic solution can be reached by numerous and sometimes very different routes. Artificial selection provides strong evidence that very similar phenotypes can emerge using entirely different genetic substrates. Two classical notions, polygeny and pleiotropy, explain the essence of this phenomenon, stressing that the majority of traits are determined by numerous genes and a single gene may affect a number of traits. Such complex interactions create robust developmental nets enabling adaptations and capable of self-regulation. Hopefully future studies will bring more understanding regarding the contribution and interaction of selection, random processes, and self-organization in the development of adaptations.

Natural selection, by creating adaptations, is supposed to drive biological systems from chaotic to more orderly states; a similar role is assigned to self-organization. Genetic drift, as an entirely random factor, should do the opposite. However, the most typical result of drift on the population level is homozygotization. This means decline in genetic variability and therefore lessening of chaos. Contrary to expectations, natural selection, particularly when it supports balanced polymorphism, may increase a proportion of heterozygotes and, if so, enhance genetic variability and hence, chaos in populations.

Summary

This chapter describes the origin, development, and incorporation of stochastic ideology into the theory of evolution. This includes consideration of genetic drift and the role of neutral mutations in evolution. The theory of genetic drift was generated by population genetics mainly due to the efforts of Sewall Wright. The evolutionary synthesis that followed used this theory for describing changes in small isolated populations, which are essential for allopatric speciation. An increasing flow of molecular data beginning in the 1960s demonstrated that selectively neutral molecular substitutions are very common. Motoo Kimura realized that stochastic factors have a much greater role in evolution than was commonly accepted and proposed the neutral theory of molecular evolution. According to these views, shared by other researchers, a significant amount of genetic variation existing in populations is purely random and does not have selective significance. Evolution of such neutral molecular variants can be described by random genetic drift. Two decades of controversy followed, which finally led to general acceptance of more moderate views and significant development of the theory of molecular evolution. As a result, randomness not only in occurrence of mutations but also in their population dynamics and eventual fixation or loss steadily became an integral part of evolutionary theory.

This certainly does not diminish the essential importance of natural selection, which operates on all levels of biological organization, including the molecular. While it is possible to successfully describe the process of natural selection using the deterministic equations of classical population genetics, there is no doubt that natural selection inevitably has probabilistic aspects. Positive selection of an allele or a trait increases the probability of its propagation but does not necessarily guarantee each individual outcome. This happens because the uncertainty of interactions of myriads of external and internal factors is so great. The universal acceptance of natural selection as the basic mechanism responsible for building adaptations should not prevent further inquiries in this matter. This includes questions about those adaptations that possibly appeared without direct involvement of natural selection. While there continue to be demands for rigorous evidence for this proposition, there are indications that random factors and self-organization may lead to adaptive changes. It is quite likely that many adaptations are caused by complex interactions of selective and nonselective factors.

References

Alstad, D.N. 2007. *Populus. Simulations of population biology*. University of Minnesota. http://www.cbs.umn.edu/populus/index.html.

Adams, M.B. 1970. Towards a synthesis: Population concepts in Russian evolutionary thought, 1925–1935. *Journal of the History of Biology* 3:107–129.

Bloom, J.D., A. Raval, and C.O. Wilke. 2007. Thermodynamics of neutral protein evolution. *Genetics* 175:255–266.

Castle, W.E. 1903. The laws of Galton and Mendel and some laws governing race improvement by selection. *Proceedings of the American Academy of Arts and Sciences* 35:233–242.

Crow, J.F. 1988. Sewall Wright (1889–1988). *Genetics* 119:1–4.

Crow, J.F. 1999. Hardy, Weinberg and language impediments. *Genetics* 152:821–825.

Crow, J.F., and M. Kimura. 1970. *An introduction to population genetic theory.* New York: Harper and Row.

Dobzhansky, Th., and S. Wright. 1941. Genetics of natural populations. V. Relations between mutation rate and accumulation of lethal in populations of *Drosophila pseudoobscura*. *Genetics* 26:23–51.

Dubinin, N.P. 1931. Genetico-automatic processes and their role in evolution. *Journal of Experimental Biology* (Russian) 7:463–479.

Dubinin, N.P., and D.D. Romashov. 1932. Die genetische struktur der Art und ihre Evolution. *Biol. Zh.* 1:52–95.

Ewens, W.J., and C.C. Heyde. 1995. Obituary: Motoo Kimura. *Journal of Applied Probability* 32:1142–1144.

Fisher, R.A. 1922. On the dominance ratio. *Proceedings of the Royal Society of Edinburgh* 42:321–341.

Fisher, R.A. 1930. *The genetical theory of natural selection.* Oxford, UK: Clarendon Press.

Haldane, J.B.S. 1927. A mathematical theory of natural and artificial selection. V. Selection and mutation. *Proceedings of the Cambridge Philosophical Society* 23:838–844.

Haldane, J.B.S. 1932. *The causes of evolution*. New York: Harper & Bros.

Hardy, G.H. 1908. Mendelian proportions in mixed populations. *Science* 28:49–50.

Heffernan, J.M., and L.M. Wahl. 2002. The effects of genetic drift in experimental evolution. *Theoretical Population Biology* 62:349–356.

Kimura, M. 1955. Solution of a process of random genetic drift with a continuous model. *Proceedings of the National Academy of Sciences USA* 41:144–150.

Kimura, M. 1962. On the probability of fixation of mutant genes in a population. *Genetics* 47:713–719.

Kimura, M. 1968. Evolutionary rate at the molecular level. *Nature* 217:624–626.

Kimura, M., and T. Otha. 1971. Protein polymorphism as a phase of molecular evolution. *Nature* 229:467–479.

King, J.L., and T.H. Jukes. 1969. Non-Darwinian evolution. *Science* 164:788–798.

Knight, C.A., N.A. Molinari, and D.A. Petrov. 2005. The large genome constraint hypothesis: Evolution, ecology and phenotype. *Annals of Botany (London)* 95(1):177–190.

Lynch, M. 2007. The evolution of genetic networks by non-adaptive processes. *Nature Reviews Genetics* 8:803–813.

Mayr, E. 1963. *Animal species and evolution*. Cambridge, MA: Harvard University Press.

Nei, M. 2005. Selectionism and neutralism in molecular evolution. *Molecular Biology and Evolution* 22:2318–2342.

Nei, M. 2008. The new mutation theory of phenotypic evolution. *Proceedings of the National Academy of Sciences USA* 104:12235–12242.

Pearson, K. 1903. Mathematical contributions to the theory of evolution. XI. On the influence of natural selection on the variability and correlation of organs. *Philosophical Transactions of the Royal Society of London, Series A.* 200:1–66.

Smith, N.G., and A. Eyre-Walker. 2002. Adaptive protein evolution in *Drosophila. Nature* 415:1022–1024.

Stern, C. 1943. The Hardy–Weinberg law. *Science* 97:137–138.

Takahata, N. 2007. Molecular clock: An *anti*-neo-Darwinian legacy. *Genetics* 176:1–6.

Tchetverikov, S.S. 1926. O nekotorykh momentakh evoliutsionnogo protsessa s tochki zrenia sovremennoi genetiki. *Zhurnal eksperimental'noi biologii,* ser. A, 2(1):3–54 (in Russian). [1961]. On certain aspects of the evolutionary process from the standpoint of modern genetics (transl. of 1926 paper). *Proceedings of the American Philosophical Society* 105:167–195.

Tihomirov, V.M. 2003. Andrei Nikolaevich Kolmogorov (1903–1987) the great Russian scientist. *The Teaching of Mathematics* 6(1):25–36.

van Nimwegen, E., J.P. Crutchfield, and M. Huynen. 1999. Neutral evolution of mutational robustness. *Proceedings of the National Academy of Sciences USA* 96:9716–9720.

Warringer, J., and A. Blomberg. 2006. Evolutionary constraints on yeast protein size. *BMC Evolutionary Biology* 6:e61.

Weinberg, W. 1908. Über den Nachweis der Vererbung beim Menschen. *Jahreshefte des Vereins für vaterländische Naturkunde in Württemberg* 64:368–382.

Williamson, S.H., M.J. Hubisz, A.G. Clark, B.A. Payseur, C.D. Bustamante, and R. Nielsen. 2007. Localizing recent adaptive evolution in the human genome. *PLoS Genetics* 3:e90.

Wright, S. 1931. Evolution in Mendelian populations. *Genetics* 16:97–159.

Yule, G.U. 1902. Mendel's laws and their probable relation to intra-racial heredity. *New Phytology* 1:193–207, 222–238.

Zuckerkandl, E., and L.B. Pauling. 1962. Molecular disease, evolution, and genetic heterogeneity. In *Horizons in Biochemistry*, ed. M. Kasha and B. Pullman, 189–225. New York: Academic Press.

chapter seven

Life

Making uncertainty certain

... scientist is possessed by the sense of universal causation...

Albert Einstein
The World As I See It
(2006)

The Cartesian Frenchman in me didn't take long to catch hold of himself and attribute those accidents to the only reasonable divinity—that is, chance.

Albert Camus
The Fall
(2004)

How to unite scientific zest for universal causation with ubiquitous randomness is a profound question. Nearly a century ago quantum physics and then genetics were forced to accept both concepts as two independent realities which seemingly contradict one another. Here we are not going deep into philosophical considerations about the essence of the material world and science. However, it is difficult to deny that randomness, or in other words uncertainty, looks like a serious challenge to universal causation and science has to moderate the importance of this traditional core principle. Surprisingly such an uneasy shift did not significantly reduce the capability of science to describe and predict. Even more surprisingly this fundamental change went relatively smoothly and did not dramatically affect the "daily business" of science. This does not mean of course that science can still operate in a purely Laplacian deterministic framework. A steady transition to a new scientific perception of the world is inevitable and ongoing.

As we saw in the previous chapters life provides endless examples of random events which can be found at all levels of biological organization. Here we shall try to understand more about transformation of this deeply entrenched uncertainty into the generally recognized "rationality" of life.

Order from chaos

Everybody knows that order is difficult to arrange and even more difficult to maintain. Because of that people often think that order, which is so typical of life, is a result of a special force, for example, natural selection. While this is correct for numerous situations, natural selection may not be the only source of order. To explore this matter a bit further, let us look at crystallization, one of the simplest instances of order in nature. There are two major stages during crystallization, nucleation and crystal growth (Wikipedia, "Crystallization"). During the first stage the molecules of the future crystals, which are dispersed in a liquid phase, start to gather into clusters. If such clusters are stable under existing conditions (temperature, supersaturation, etc.) they become nuclei for the subsequent crystal growth. Supersaturation is the driving force of the process and, once it is exhausted, the solid-liquid system reaches equilibrium and the crystallization is complete. A crystal is more easily destroyed than it is formed. As experience shows it is usually less difficult to dissolve a perfect crystal than to grow again a good crystal from the resulting solution. The nucleation and growth of a crystal are under kinetic, rather than thermodynamic, control. The kinetic factor itself is affected by random events. Entropy decreases during the crystallization process, which occurs due to ordering of the molecules within the system, and is compensated by the thermal randomization of the surroundings, due to the release of the heat of fusion (Wikipedia, "Crystallization").

Some conclusions can be drawn from this example. First, self-organization does occur as a consequence of a certain balance between kinetic and thermodynamic forces. Second there is a scope for random events in such an orderly process as crystallization. With some caveats this logic can be spread to other more complex interactions between heterogeneous molecules, which may also lead to the emergence of complex molecular structures. The contribution of self-organization to the formation of numerous biological structures remains an open question. It is clear, however, that some elements of order, which we observe in living creatures, do emerge on the basis of simple molecular interactions. Such interactions probably were profoundly important during the prebiotic stage, which eventually led to the origin of life on our planet. We shall discuss this question later in the chapter.

The emergence of order from chaos also has been described in mathematics. The quadratic iterator

$$x_{n+1} \to rx_n(1 - x_n),\ n = 0, 1, 2, \ldots$$

where x_n is a number between zero and one and r is a positive number, demonstrates that alternative states can be ruled by a single law. Study of this iterator provides a vivid example of order formation from chaos, when r has certain values (Figure 7.1). In 1975 M. Feigenbaum discovered that with r value between 1 and 3, x will stabilize on the value $(r-1)/r$. Between 3 and $1+\sqrt{6}$ (approximately 3.45) oscillations between two x values may occur forever. When r values are between 3.45 and 3.54 (approximately) oscillations occur between four values. If r values are growing further, oscillations will occur between 8 values, 16, 32, etc. "The ratio of the difference between the values at which such successive period-doubling bifurcation occurs tends to a constant of around 4.6692 … He (*Feigenbaum, AR*) was then able to provide a mathematical proof of this fact, and he then showed that the same behavior, with the same mathematical constant, would occur within a wide class of mathematical functions, prior to the onset of chaos. This universal result enabled mathematicians to take their first steps to unravelling the apparently intractable 'random' behavior of chaotic systems" (Peitgen, Jurgens, and Saupe 2004). This "ratio of convergence" is now known as the first Feigenbaum constant (http://en.wikipedia.org/wiki/Logistic_map). This mathematical illustration shows that chaotic behavior may arise from a simple nonlinear dynamic equation and that a chaotic system may become orderly due to its intrinsic properties (see Peitgen, Jürgens, and Saupe 2004).

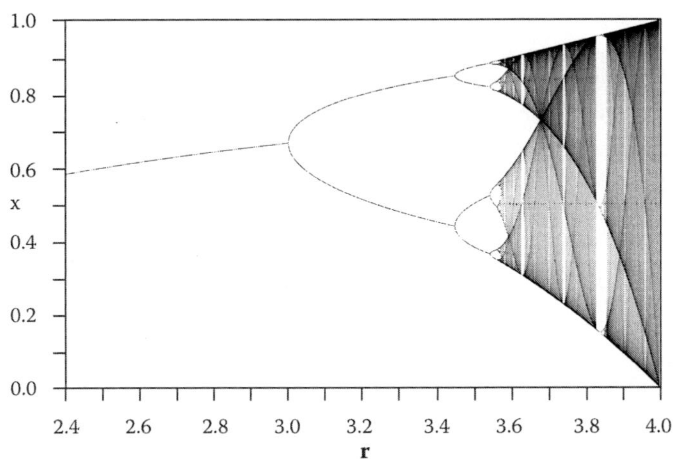

Figure 7.1 A bifurcation (Feigenbaum) diagram graphically representing the results of the study of the quadratic iterator $x_{n+1} \rightarrow rxn\ (1-xn)$, $n = 0, 1, 2, \ldots$ where x_n is a number between zero and one and r is a positive number. This iterator provides an example of order formation from chaos, when r has certain values (see text for further explanations). (From Wikipedia, http://en.wikipedia.org/wiki/Logistic_map.)

Thus, physics and mathematics provide solid evidence that transitions between chaos and order are possible and they do not require a specific natural force that is aimed toward a particular outcome. There is no reason to believe that biological systems, despite their complexity, are exempt from these general rules. On the contrary self-organization is frequently observed at different levels of biological organization (Camazine et al. 2003). Formation of complex patterns in biological systems, or one would say order, results from iteration of simple elements. This is rather similar to observations made in physics and mathematics.

The question of whether self-organized patterns have been always molded by natural selections remains contentious. Camazine et al. (2003) as well as many biologists believe that this is the case and the alternative view is a serious misunderstanding. They think that the ground work for self-organization had been always done by the preceding natural selection. To provide concrete evidence for every case of self-organization is a nearly impossible task and even when done, this opinion rests partially on belief in the complete universality of natural selection. Kauffman (1995) and other researchers, on the contrary, argue that natural selection is hardly the sole source of order in living nature. This certainly is not equal to rejection of natural selection as a decisive force of biological evolution. Perhaps the origin of life is the area where the question of self-organization and natural selection could be tested more convincingly. It is a quite common view that natural selection, at least in its classical form, could not operate outside the organic world. If this is the case, one may guess that self-organization and random molecular events were the only driving forces that eventually led to the origin of life. A logical dilemma from such a conclusion is obvious; if self-organization played such a pivotal role in the origin of life why should it disappear entirely from the acting factors after the onset of life?

Perhaps the story is not as straightforward as described above and some kind of "molecular selection" that transformed protometabolism to metabolism might be operating even at a prebiotic stage (de Duve 2007). The most recent attempt to deduce what might have happened prior to the origin of life indicates that prelife or chemical evolution could have had "mutations" and "selection" without replication (Nowak and Ohtsuki 2008). According to these views "prolife is a scaffold that builds life" and competition between prelife and life took place. As soon as efficiency of replication exceeded a critical level, life outcompeted prelife. This phase transition might have been very fast. If such a scenario is true, the common view that replication is a prerequisite of selection is not necessarily valid but instead could be a selection for replication.

Regardless of the scenario self-organization should not be excluded from the picture. It appears to be possible that the role of self-organization was somehow limited when natural selection steadily became the dominant driving force of evolution and the outcomes of natural selection and self-organization intertwined. However, the view that natural selection and self-organization should be seen as competing ideas is too simplistic (Richardson 2001).

Orgel (2000) expressed doubts that complex organic molecules, which were necessary for creation of sufficiently complicated organization and capable of crossing the border between nonlife and life, "might be made available in large amounts via self-organizing." Rather, the "autocatalytic cycle might, in principle, help to explain the origin of the component monomers." Earlier Eigen and Schuster (1977), while trying to find a solution to the paradox that no enzymes are possible without a sufficiently large genome and large genomes are not possible without enzymes, proposed the idea of a hypercycle. The latter is a closed cycle of simple replicators, possibly RNA molecules, whose replications might be affected by the preceding reactions in the hypercycle and might influence the following steps. Although this idea seems to be relevant to a more advanced stage of biogenesis, it provides a very useful concept of consequential development of more complex systems from simple systems. Self-organization of such hypercycles with or without "molecular selection" had to be a critically important requirement; otherwise, durability of these emerging blocks of life would be too low. In any case the hypercycles are considered a very fruitful idea, which can provide another illustration of creating order from chaos.

The very early stages of transition to life must have been based on chemical processes that could operate continuously for a long time, change conditions in a particular area, and thus fulfill necessary prerequisites. In some way these processes should resist overwhelming randomness, which always increases entropy. Kauffman (1993) suggested that autocatalytic processes self-arranged into circuited systems were likely candidates for such processes. Investigations of such autocatalytic processes can provide useful information about possible intermediate reactions on a pathway from the chemical to the biological world. For instance, the Formosa reaction, which is an autocatalytic aqueous synthesis of monosaccharides from formaldehyde in the presence of calcium hydroxide, is considered one of the possible steps in this direction. This reaction possesses some "mutation" properties and may eventually lead to production of ribose, which is a basic component of RNA molecules (Parmon 2008).

There is another interesting argument in favor of hypercycles. Lee et al. (1997) reported "a chemical system that constitutes a clear example of a minimal hypercyclic network, in which two otherwise competitive self-replicating peptides symbiotically catalyse each others' production." This could be a demonstration of a transition from a complex chemical reaction toward "living chemistry." Self-organization of autocatalytic reactions, at least during the very early stages of transition to life, was most likely dictated by physical and chemical parameters. A critically important condition for natural selection, namely the existence of a proper "hereditary" memory, was not achieved as yet. Despite all limitations and probably fragility of prebiotic "living chemistry" systems they should not be considered as entirely accidental. Their existence was warranted by fundamental physical constants as well as kinetic and thermodynamic laws. The most exciting part of the process must be the ability of such systems to evolve and became more complex.

Complexity is one of the obvious and fundamental characteristics of life. Since Darwin the question of why evolution usually, but not always, leads to an increased complexity has remained open for exploration (Zuckerkandl 2005). It seems to be a commonly held view that biological complexity is a result of countless interactions of networks on cellular and organismal levels (including genetic, biochemical, developmental, and morphogenetic networks), as well as at the level of populations, species, and ecosystems. As has been shown recently, the underlying genetic networks appear "to operate either in the ordered regime or at the border between order and chaos but does not appear to be chaotic" (Shmulevich, Kauffman, and Aldana 2005). This quasi-ordered complexity of living organisms, which is probably a combined outcome of natural selection and self-organization, is the filter for omnipresent random events occurring with an incredible regularity and rate. Transformation of randomness into a complex and adaptive biological organization is the very core of evolutionary processes.

Perhaps principles of self-organization and selection have more universal value than usually appreciated and spread beyond life phenomenon. The Internet is a well-known example of an evolving and chaotic system. Self-organization, as has been shown, is active on the Internet (Barabási 2003). Selection in favor of more popular web sites is also supported by observations. Such sites grow faster and the number of those covering the most popular topics increases.

What is life?

This question has been raised many times. Erwin Schrödinger, as mentioned in Chapter 2, succeeded in answering the question more than others. Schrödinger tried to develop his understanding from a physical point of view. Some critics of this approach, including such outstanding biologists as Ernest Mayr (2004), believed that the irreducible complexity of organisms did not make the application of physical principles to biology feasible. If that is true then one would expect that at least some physical laws do not work in biological systems. So far no one has been able to prove such a statement and this includes the first two laws of thermodynamics. Despite the fact that life forms have a number of specific features, which are not known for other forms of matter, they do not deviate from any known physical principle.

In the previous section we have already discussed complexity as an important characteristic of all living beings. Two other features, heterogeneity and compartmentalization, are tightly linked to complexity. Effectively organisms are complex because they are composed of many different elements (heterogeneity) and components of cells and individuals are separated one from another (compartmentalization). The degree of complexity, heterogeneity, and compartmentalization existing in biological systems is exceptional but these characteristics can also be found in abiotic nature. A fairly complex rock, with significant heterogeneity and clear compartmentalization, is a good example. Thus, explicit description of life using these important characteristics is hardly sufficient.

Three other features, namely reproduction, adaptability, and genetic variability, are much more life-specific and also firmly interconnected. We saw that genetic variation, which is the prerequisite for any adaptation, usually occurs during reproduction. And vice versa, without ongoing variability and adaptation long-term reproduction is unlikely. If not all then at least many attempts to describe life comprise these three core features. Francis Crick (1988), for instance, stressed the role of reproduction and adaptation: "Outside biology, we do not see the process of exact geometrical replication…" and further "What gives biology its special flavor is the long-continued operation of natural selection."

Defining the most characteristic features of life does not necessarily bring us closer to answering the question about the physical essence of life. Two explanations are commonly used to elaborate on this question. One of them considers reproduction of genes and hence replication of DNA molecules as the center point of life. While this process is indisputably critical for all known independent forms of life, it is unlikely it was so important in the early stages of biogenesis prior to the origin of life proper, when genes had not been invented as yet. The alternative energy-related explanation seems to be more fundamental. It is common knowledge that every species tends to increase energy consumption and only physical and ecological factors limit this perpetual trend. Reproduction is rather the means for achieving this energy objective than the objective itself. As soon as a species enters a niche, which is relatively free of limiting factors, it begins nearly exponential reproduction. Speaking metaphorically a species consumes energy and matter like a black hole, with a major difference that both energy and matter easily escape from a biological system. One could say that life is a very "aggressive" form of matter which, however, can exist only in a narrow range of physical conditions. Energy flow through a biological system keeps it far from thermodynamic equilibrium, which results in an increasing order and hence lower entropy. It was understood long ago that the increase is not in defiance of the second law of thermodynamics simply due to the fact that all biological systems are open. Increasing order is possible because life is based on biochemical reactions with extremely biased equilibriums which are consequences of catalytic activities of numerous enzymes.

It should not surprise us that energy, which is the core physical concept of the universe, is also the major driving force and "objective" of life. One should keep in mind that the phenomenon of life, despite its probabilistic nature, is a consequence and continuation of evolution of the universe. Morowitz and Smith (2007) expressed their support of the idea that life requires a source of free energy and mechanisms necessary to capture and utilize it. They also presume that the reverse could be true and "the continuous generation of sources of free energy by abiotic processes may have forced life into existence as a means to alleviate the buildup of free energy stresses."

The old comparison: physics and biology

It is no secret that physics and biology are different. No wonder; they explore different realities. Physics attempts to understand the whole universe and biology makes an effort to understand a tiny but very special and terribly complex section of the material world. A number of physical principles have been steadily discovered during the last three centuries, which have universal value and are generally relevant to the majority of, if not all, known events and processes. It probably would be fair to say that these are laws of nature discovered by physics, rather than physical laws, as they are applicable to all other areas of science. Other scientific disciplines like chemistry, biology, and geology also formulated certain rules and laws; however, by necessity they cover only a part of everything.

Even more importantly, the boundless variation of biological systems is a significant obstacle for generalizations that can be applicable to all biological objects and processes. There are only two exceptions which traditionally were not recognized as general biological laws. Both were discovered by Charles Darwin and co-authored by Alfred Wallace. Certainly these are genetic variation and natural selection. Perhaps verbal formulation of these basic principles was not sufficient to coin terms like the first and the second Darwinian laws. Nevertheless even when the principle of natural selection was reformulated in mathematical terms by Ronald Fisher (1930) and named "the fundamental theorem of natural selection" this did not change traditional perceptions. It is correct that the theorem covers only some population genetics situations, but it seems likely that a more general theorem that would be able to cover all circumstances hardly would have much warmer reception. There are a number of reasons for this. First, many biologists do not feel that such formalization would make a big difference. Second, the majority of scientists, and not without a good reason, believe that this principle has special biological relevance and is unlikely to be important as a universal principle. Whether the latter is correct is not absolutely clear and we return to this matter in the following section.

Despite the objective difficulties in developing theoretical biology, many physicists permanently or temporarily moved to biology and made a tremendous contribution. Max Delbrück and Francis Crick are among the most well-known "converts." There are many other researchers who were spectacularly successful in such a transition and several of them received Nobel prizes. In some cases the motivation was to discover general principles by studying biological processes. More typically the intention was to build theoretical biology and describe biological processes in physical terms. However, as Rashevsky (1966) wrote:

> It must be kept in mind that representing individual biological phenomena in terms of physics is not the same as deducing from known physical laws the necessity of biological phenomena. Drawing an analogy from pure mathematics, it is possible that while every biological phenomenon may be represented in terms of physics, yet biological statements represent a class of "undecidable" statements within the framework of physics.

Both physics and biology independently came to the realization that the behavior of so-called simple systems, like an ideal gas or Mendelian populations, cannot be described using a "bottom up" approach. In other words a description of an ideal gas must be done using integrative parameters such as energy, temperature, pressure, and entropy. Similarly, a description of Mendelian populations must be made using parameters such as allele frequencies, selection, heterozygosity, and polymorphism. This clearly shows that the behavior of even simple systems cannot be described using information only about their basic components. The reason for this is known; complete information about all elements comprising a simple system is not available either due to intrinsic causes (uncertainty and randomness), and in this case principally cannot be obtained, or collecting such information is practically impossible. The perennial question why in complex structures the sum of elements is not equal to the whole system has a solution if we accept that our knowledge about the elements and their interactions is never complete. There is also another side of this story. This phenomenon, more recently tagged as "emergence," has been well known for a long time in different fields of science. Let us consider here a biological example; individual neurons are unable to think but the human brain, which is composed of many neurons, surely can. It means that connections between many neurons are critically important for producing the quality which does not exist in individual elements. This is a good demonstration of a familiar dialectical statement made long ago by German philosopher Hegel (1874) concerning the transition from quantity to quality. Life is an endless chain of "emergences" and because neither of them, as far as I am aware, has a clear mathematical or physical interpretation, it is difficult to apply formal methods in biology.

The problem whether energy could be used as a parameter that can unify nonbiological and biological systems still awaits resolution. The work of Sella and Hirsh (2005) gives an indication that it might be achievable. These authors refer to Ronald Fisher (1930), who wrote that "the fundamental theorem [of natural selection] bears some remarkable resemblance to the second law of thermodynamics. It is possible that both may ultimately be absorbed by some more general principle." The parallels between statistical physics and evolutionary dynamics, which these authors emphasized, "permits derivation of an energy function for evolutionary dynamics of a finite population." As Sella and Hirsh (2005) claimed, "the form of this energy function is precisely that of free energy, and the maximization of free fitness is precisely analogous to the second law of thermodynamics." Statements of this kind provide support for cautious optimism that certain biological phenomena might be eventually defined in physical terms. Attempts were also made to model animal growth and production efficiency using parameters of the internal combustion engine, which in turn can be described in physical terms (Kinghorn 1985).

How far these analogies can go is a hard question. Several factors seem to be influential. As has been shown in the previous chapters, living matter is dramatically affected by randomness. Tossing a coin is a classic example of a random process; the important idea is that any toss of the coin is not affected by the outcome of the previous toss. An atom or a molecule may lose an electron and then regain it back and these events are completely reversible. In biology random events that happened in the past usually affect the future. For instance, if a population was affected by random forces and as a result the frequencies of alleles were changed, this will influence the frequencies of alleles in the following generations and likely the fate of this population is going to be different. Changes of this kind cannot be reversed because biological systems are too complex and the probabilities of such events are close to zero. Does this mean that if the evolutionary tape was replayed, as S.J. Gould (1989) expressed this idea, the results would be different each time?

We discussed a similar issue in Chapter 4 in relation to individual development and came to the conclusion that most likely each realization of the same genetic program in identical conditions will not produce individuals with identical phenotype. Then why should one expect that the outcome of the entire evolutionary process would be the same, if it could be "replayed" several times? What we do not know is how far the outcomes of imaginary evolutionary processes would deviate. Obviously this type of uncertainty does not fit well into a framework of a precise science. As Francis Crick (1988) brilliantly put it: "What seems to physicists to be a hopelessly complicated process may have been what nature found simplest, because nature could only build on what was already there." Similar factors are also at work in other than life forms of matter but their influence is much less profound.

No doubt physical laws operate in biological systems and possibly in the future the energy principle or other theoretical ideas could bring different fields of science, like physics and biology, closer. One could also question whether a reverse flow of ideas might occur from biology to other fields of science like physics or chemistry. For the reasons discussed in this section such a possibility looks remote. However, the principle of natural selection "born and bred" in biology could be an exception and its universal value is not out of the question.

Natural selection: biology and beyond

The sole fact that rigorous and wide discussion about natural selection has continued for the last 150 years confirms the paramount importance of this concept not only for evolutionary biology but also for other fields of human knowledge. Every diligent student knows that a scientifically sound explanation for the countless adaptations ever developed by life forms is natural selection. While there are several other factors fueling biological evolution, selection is the one capable of producing adaptations. This whole principle is based on precedents. As soon as an individual possessing a set of ordinary or unusual traits is capable of surviving and reproducing in existing conditions, it means that the individual has passed a test of natural selection and his genes will be represented in the next generation. The fate of all individuals and their genes is always under tough scrutiny, particularly if conditions change. Francis Crick (1988) made a dazzling comment about the origin of molecular adaptations: "If this were produced by chance alone, without the aid of natural selection, it would be regarded as almost infinitely improbable."

Despite the great importance of natural selection, a large fraction of neutral or even slightly deleterious (near-neutral) mutations become fixed in populations by genetic drift. A recent revision of the role of adaptive near-neutral and neutral mutations confirms not only differences in predictions made by selection and drift hypotheses but also the fact that selective and random factors are quite compatible during evolution (Ohta 2002). Natural selection and random forces, including the mutation process, recombination, and genetic drift, are independent factors of the evolutionary process. Their contribution is influenced by the population size, stability of environmental conditions, and a number of other parameters. Selection and randomness interact in many ways. For instance, there are mutations that alter the noise of gene expression (for details see Chapter 4). The existence of such mutations "suggests that noise is an evolvable trait that can be optimized to balance fidelity and diversity in eukaryotic gene expression" (Raser and O'Shea 2004). In other words selective factors are capable of optimizing the level of stochasticity in evolving biological systems depending on changing conditions. This is a very complicated system of interactions, indeed.

A distinction between what is a subject of selection and what is not could be in some instances quite unclear. It is well known that genomes are full of DNA sequences that are noncoding at a particular level. Only about 5% of mammalian genomes is under selection and protein coding genes constitute approximately 1.5% of the total. For instance, introns are not directly involved in protein coding and in this sense do not have a selective value. However, introns may influence chromatin structure or affect RNA splicing, which is very much relevant to protein synthesis; hence, they have some selective value. Zuckerkandal (2002) wrote in this regard: "a sequence that is non-functional at one level of nucleotide plurality may participate in a functional sequence at a more inclusive level." This is another indication of the complexity of life and the process of natural selection. The problem of neutral or selective values of the same section of DNA under variable circumstances is also very much pertinent here.

A number of concerns were expressed about indiscriminate use of the principle of natural selection as an explanation for biological phenomena without clear evidence. A central message of these concerns is correct; a scientific proof should not be replaced by belief in the correctness of a general idea. On the other hand hypothetic explanations based on the principle of natural selection can be used until tested and proven wrong. Hence, the key question is whether a reliable test is always possible; unfortunately this is not as easy as one may expect.

During the last two decades Daniel Dennett emerged as the leading philosophical proponent of the principle of natural selection. The Darwinian idea, as he suggests, "is the best idea anybody ever had, ahead of Newton, ahead of Einstein. What it does is it promises to unite the two most disparate features of all of reality. On one side, purposeless matter and motion, jostling particles; on the other side, meaning, purpose, design" (Interview with Daniel Dennett, http://www.pbs.org/wgbh/evolution/library/08/1/text_pop/l_081_05.html).

Dennett (1995) in his book *Darwin's Dangerous Idea* brings together science and philosophy and put the Darwinian idea in the center of philosophical considerations. In the book he moves beyond numerous and witty considerations of biology proper. One such intellectual pursuit is the consideration of Richard Dawkins' (1976) idea and term *meme*, which refers to any thought or behavior that can propagate in human societies by means of imitation. Some memes come and go; others stay for a very long time. There are memes that quickly become universal and there are those that never spread beyond a small group of people. The dynamics, or one could say evolution, of memes shows a significant similarity with biological evolution. To reach the point Dennett (1995) accurately outlines the basic principles of evolution by natural selection, including variation, heredity or self-replication, and differential fitness. Then he states that these principles "though drawn from biology" do not necessarily refer to molecules, genes, or life itself. From this observation Dennett came to the conclusion that meme evolution "obeys the law of natural selection quite exactly," a very interesting conclusion that effectively means that the Darwinian principle has broader implication than just for biological evolution. How much broader remains to be seen.

Earlier we touched briefly on the origin of life. Among many serious difficulties in understanding what happened on our planet more than three and a half billion years ago, there is the problem of natural selection. One of the basic requirements for natural selection to be an acting force is the existence of a hereditary mechanism as simple as possible but nevertheless available. However, before sufficiently complex hereditary molecules were invented, such a mechanism hardly could exist. This looks like the problem of the chicken and the egg. Perhaps a way to resolve this impasse is to presume that more simple forms of selection might guide the processes of prebiotic or in other words chemical evolution. This would entail a significant relaxation of the meaning of selection. Such a process could hardly be named natural selection in the Darwinian sense.

de Duve (2007) wrote about molecular selection, which could have "played a role at the dawn of the RNA era," but he also mentioned that such a role can only have been limited. In much earlier stages of evolution even this type of selection was less prominent. Then, autocatalytic reactions on the surfaces of certain inorganic substances could support sufficiently stable production of the basic components of future life (Parmon 2008). There is nothing teleological in this concept if we appreciate this chemical build up toward the following biogenesis only from the point of view of the successful origin of life. In that regard such reactions were "selected." Otherwise they probably were just the most stable and durable chemical reactions due to their kinetics and thermodynamics. This kind of "selection" of the most stable and durable reactions might be not very far different from the self-organization of autocatalytic processes suggested by Kauffman (1993). This idea looks a bit hypothetical but, if the classical form of natural selection evolved from earlier forms of "chemical" or "molecular" selection, the roots of natural selection might be in self-organization, the intrinsic property of matter.

The principle of natural selection has also promoted quite unusual ideas in cosmology. In the late 1980s Lee Smolin (1997) put forward a hypothesis he called cosmological natural selection. Smolin (2006) confessed that in order to resolve a problem of multiple universes (a multiverse theory) he adopted the idea and the term natural selection. It is hardly appropriate to discuss here the essence of the multiverse theory and I refer those who might be interested to Smolin's books. The point I try to emphasize, however, seems quite obvious. The idea of natural selection is so powerful that it spreads into different fields of science and generally human knowledge. Another potent demonstration comes from artificial intelligence studies, a fast developing field of computer science. An evolutionary algorithm is a method of computations that emulates evolution or, better to say, population genetic processes by using the ideas of reproduction, mutation, recombination, and selection. Candidate solutions of a problem reproduce, mutate, recombine, and compete. As a result the best solutions can be successfully and efficiently found in such diverse fields of human endeavor as engineering, art, biology, economics, robotics, social sciences, physics, chemistry, and politics ("Evolutionary algorithms," Wikipedia, http://en.wikipedia.org/wiki/Evolutionary_algorithm). In many cases the evolutionary algorithms are the only means of solving highly complex problems.

As mentioned earlier the Internet also provides numerous examples of selection: more popular web sites grow faster and their number increases. A conclusion can be drawn that the principle of natural selection has general importance and perhaps one day may join other basic laws of nature. If this ever happens it will be the only general principle of nature initially discovered by biology.

Randomness: nuisance or essence?

Now once again we return to randomness, which is the cause of genetic variation and evolutionary novelties. Earlier in Chapter 5 we identified three distinct sources of randomness operating in living organisms: quantum uncertainty, stochasticity operating on different levels, and finally "organized randomness" related to allele transmission. Quantum uncertainty and stochasticity are fundamental characteristics of matter that existed prior to the origin of life. Their importance was steadily accepted as physics progressed in the nineteenth and particularly the twentieth centuries. This was not an easy shift in understanding nature and there were significant disagreements. The refusal by Albert Einstein to accept quantum theory and his views on the nature of stochasticity, clearly expressed in the famous remark "God does not play dice," are the best indications of how difficult it was for him to give up the idea of universal causation. Einstein could not agree that complete knowledge was not only technically but even theoretically impossible. Stephen Hawking (Public lecture, http://www.hawking.org.uk/lectures/dice.html), on the contrary, recently stated that "Not only does God definitely play dice, but He sometimes confuses us by throwing them where they can't be seen." And he continued: "Many scientists are like Einstein, in that they have a deep emotional attachment to determinism." So it is too early to proclaim that a unanimous point of view has prevailed and it is crystal clear for everybody that randomness is not a nuisance but rather the essence of natural processes.

Life is particularly dependent on randomness. There is no doubt that sexual reproduction and transition to multicellularity, as discussed in Chapter 5, increased usage of randomness very significantly. The emergence of meiotic chromosome segregation and random fertilization were crucial in promoting this organized type of randomness. These facts alone justify the conclusion that randomness was in high demand as biological evolution progressed. One should keep in mind that, beyond these types of randomness leading to genetic variability, there are also many biological phenomena that are based on random events. A substantial analysis of cell stochasticity related to developmental processes was compiled by Theise (2006). An example of randomness discovered in Yanagida's laboratory is impressive (Ishii, Nishiyama, and Yanagida 2004). This study shows that the movements of a single myosin filament along a single actin filament, which are the basic components of muscle cells allowing contraction and relaxation, are random and promoted by thermal fluctuation, or Brownian motion of water molecules. Traditional views that the hydrolysis of ATP molecules provides enough energy for such movements is incomplete; rather this energy is sufficient for constraining the random movements into physiologically required acts. The authors concluded that "protein interactions bias the random thermal noise in a manner such that the protein can perform its given functions" (Ishii et al. 2008). This is a quite remarkable demonstration of the transition from chaos to order and additional proof that characterization of complex biological systems usually does not fit either the rigid determinism framework or complete randomness; instead they can be described as systems with constrained randomness (Theise and Harris 2006). Also there is a large set of biological phenomena above the individual level which demonstrate transition from disorder to order. It was shown that "as the density of animals in the group increases, a rapid transition occurs from disordered movement of individuals within the group to highly aligned collective motion" (Buhl et al. 2006).

Cells as fundamental blocks of life show considerable physiological plasticity, which is important for cell-to-cell interactions and hence for normal development and functioning of tissues, organs, and the whole organism. This is poorly consistent with the deterministic vision of developmental processes. Without negating the importance of a genetic program providing the direction for individual development, the role of stochasticity in gene expression, cell differentiation, and interactions is reevaluated in recent publications. Some genes that are expressed stochastically in isogenic cell populations and different cells within a population have variable threshold to external stimuli. It has been proposed that self-organization and stochasticity at molecular, cellular, and possibly higher levels of organization could be used as a unifying conceptual framework rather than a nuisance that spoils the perfect deterministic vision of development. Self-organization through stochasticity becomes a core topic in such a discussion (Kurakin 2005; Theise and Harris 2006).

Using stochastic ideology is not an entirely new approach in genetics. Quantitative genetics, since its inception in the early 1920s, effectively removed genes from direct consideration and instead applied the power of statistical methods in order to estimate phenotypic and genotypic variances, heritability, estimated breeding values, response to selection, and other relevant parameters. It might look surprising but in many cases attempts to exploit direct knowledge about genotypes were not as successful as the statistical approach used by quantitative genetics. Population genetics, as discussed in Chapter 6, also incorporated stochastic methodology long ago, when the idea of genetic drift was initially proposed. A similar logic has eventually led to the theory of molecular evolution. It seems that further penetration of the stochastic approach is inevitable and not only as the method for description of biological reality but also as the core philosophical principle. In this newly emerging framework randomness is definitely the essence rather than a nuisance.

The reason and the consequence

Hardly anyone raised this issue more profoundly than Jacques Monod (Figure 7.2 and Box 7.1) in his book *Chance and Necessity* (1971). As the blurb on the front jacket indicates, Monod's intention was to develop "a philosophy of a universe without causality." This is no small task by any standard! The book begins with a citation of Democritus: "Everything existing in the Universe is the fruit of chance and of necessity." Despite such early recognition of the importance of chance in the history of human thought, science only began appreciating this fundamental aspect of nature not very long ago. Monod stated in the book "Chance alone is at the source of all novelty, all creation in the biosphere."

Monod canvassed his vision of biology among other branches of science. He believed that biology has both marginal and central roles. The marginal role of the living world is the result of its tiny size in the universe. The central role of biology is related to man's relationship to the

Figure 7.2 Jacques Lucien Monod (1910–1976), outstanding French biologist and Nobel laureate (1965). Monod made important contributions to enzymology and allosteric regulation. Together with François Jacob he developed the idea of the operon. Monod contributed to the philosophy of science. He was born in Paris and died in Cannes, France. (Courtesy of ASM Press; from Origins of Molecular Biology: A Tribute to Jacques Monod, ed. Agnes Ullmann. Washington, D.C.: ASM Press, 2003.)

universe. Biology goes directly to the core problems that must be resolved before understanding the role of the human mind in developing knowledge about the universe. Human capacity to reflect nature is affected by human logic and sensory capabilities. How much these limitations might influence our understanding of the world is a complicated matter.

The basic tenet traditionally used by science is establishment of cause-effect relations. Einstein, as follows from the epigraph to this chapter, emphasized the importance of universal causation as the leading scientific principle. Perhaps the majority of scientists try to find cause-effect relationships every day. Unfortunately random events

BOX 7.1 JACQUES LUCIEN MONOD

Jacques Lucien Monod was born in Paris in 1910. His father was a descendant of a Swiss Huguenot family and his mother was American born, in Milwaukee. Monod's parents did their best to provide a good education and encourage broad interests in music, arts, and science. Seven years later the family moved to Cannes in the South of France. Monod learned to play the cello very early and he remained a passionate musician for the rest of his life. His interest in biology grew steadily, including during his years at the College de Cannes, from which he graduated in 1928. Monod received a B.S. from the Faculte des Sciences at the University of Paris, Sorbonne, in 1931. He then worked for a short time at Roscoff marine biology station which was a very useful experience. There Monod met André Lwoff, a microbiologist with whom he established a life-long collaboration. Another outstanding biologist, Boris Ephrussi, working at Roscoff shared his knowledge in physiological and biochemical genetics. Soon after this Monod took up a fellowship at the University of Strasbourg. In 1932 he won a Commercy Scholarship and returned to the Sorbonne as an assistant in the Laboratory of the Evolution of Organic Life. In a short while Monod became an assistant professor of zoology. In 1936 using a Rockefeller grant Monod and Ephrussi went to the United States to the California Institute of Technology, where he studied the genetics of the fruit fly (*Drosophila melanogaster*) under the direction of Thomas Hunt Morgan. This was a great experience which strongly influenced the young researcher. After returning to France Monod completed his studies and obtained a Ph.D. in 1940. During World War II Monod joined the French resistance movement, where he played an active role. Starting in 1944 he worked at Lwoff's laboratory. Monod continued his studies on bacterial growth, which could be described in a simple, quantitative way and guessed that the bacteria had to employ different enzymes to metabolize different kinds of sugars. He was particularly interested in a form of "enzyme adaptation," when a colony switches between enzymes depending on their presence in the substrate. Steadily he began investigations of genetic consequences of the growth patterns and the enzyme induction, which were published in 1953. The same year Monod became the director of the department of cellular biology at the Pasteur Institute and commenced his collaboration with François Jacob. Together they investigated the relationship between heredity and environment in enzyme synthesis, which eventually led to their famous model of regulation of protein synthesis in bacteria based on the idea of the operon. The importance of these results became obvious rather soon and Monod was invited to take a concomitant position of professor of biochemistry at the Sorbonne. In 1965 the Nobel Prize was awarded to F. Jacob, A. Lwoff, and J. Monod for "discoveries concerning the genetic regulation of enzyme and virus synthesis." During the following ten years Monod sharply shifted his activities to administrative duties, particularly after 1971, when he became the director of the Pasteur Institute. In 1970 he published a small but very influential book, whose English translation Chance and Necessity in 1971 became a bestseller. Soon after the death of his wife in 1972 Monod fell ill and then four years later died at his home in Cannes, in his beloved South of France.

Sources: http://www.faqs.org/health/bios/60/Jacques-Lucien-Monod.html; http://nobelprize.org/nobel_prizes/medicine/laureates/1965/monod-bio.html.

break the chain of causal connections and diminish the universality of this principle. Questions such as what is the cause for this mutation, recombination, or segregation event do not have answers. Science may explain how this did happen rather then why. It means that there is a limit to understanding causes of life phenomena. If so, discovering "universal causation" in biology and probably in other branches of science is a futile task. This general limitation does not prevent predictions; however, their accuracy usually diminishes in time. Life, as well as evolution of the universe, is not completely determined by the laws of nature, as Laplace believed.

Monod (1971) stressed that life is a unique occurrence that cannot be deduced from the very basic principles. By making such a statement he did not deny that life phenomena can be explained through application of the same principles. This reveals logical asymmetry which is the result of omnipresent randomness and uncertainty. Let us consider a simple example in order to clarify the problem. If we disband a rather simple technical devise there is a pretty good guarantee that after some attempts we shall be able to reassemble it to the initial state. Could we do the same with the most primitive living being? The answer is certainly "no" and not only because the complexity of life phenomena is very high and our skills and knowledge are insufficient. Even if we could reassemble a living being it would never be exactly the same. The major problem of this imaginary task is the always existing break in the chain of causal connections which in principle prevents complete understanding of a biological object. Obviously without complete understanding reassembly of a living organism is not feasible. A similar logic is also correct for nonliving objects.

The only significant difference is a degree of complexity and randomness, which is much higher in life forms.

A genome of a cell, as is usually said, contains all the hereditary information about the cell. Does this mean that in a thought experiment this information is sufficient for creating the same cell? Again the answer is "no" and mainly because the total volume of information contained in a cell is larger than what is contained in the genome. The additional information, which is not stored in the genome, was accumulated during the previous cell cycles and affects numerous features of a cell. The volume of this additional information becomes even larger as soon as we move from a cell to an organism. Essentially this additional information carries traces of numerous random molecular interactions and environmental influences which in some degree define the current status of a cell or an organism. Such information, which can never be measured, in part may explain the deficiency of the reductionist approach. However, the reductionist approach is the only one that allows understanding of how living beings are constructed and how they function. Without this often criticized but imperatively important method progress in understanding of life, and matter in general, is nearly impossible.

The existence of this hidden information is the major stumbling block in the way of a deterministic explanation of nature. And here there is a split between two alternative points of view, deterministic and indeterministic. In a very condensed form this dilemma could be defined as whether the causes of randomness can ever be understood. In short the deterministic camp assumes that randomness is what has not been studied properly as yet and this should happen in the future. The indeterminists argue that science has reached the limit and further understanding of sources of randomness is impossible, not only because we cannot "get there" but because "there is nothing there." In other words, some questions do not have answers in principle.

From this point we can move toward philosophic interpretations of these meanings. Stamos (2001) made a thorough contribution to this discussion in his review. Among other useful ideas he gave a clear and succinct definition of determinism: determinism = [every event requires a cause] + [same cause, same effect]. These two deterministic principles were and to some degree still are the "α and ω" of the scientific approach. However, during the twentieth century, there were many major shifts in scientific methodology and mentality. Physics led the charge. Biology, and genetics in particular, offered numerous facts and ideas raising significant doubts about the logical predicate "every event has a cause." And furthermore, a lack of the cause at least for some events devalues relevance of the second deterministic principle. Many of these ideas and facts, presented in the previous chapters of the book, show that indeterminism, promoted by ubiquitous random events at all levels of biological organization, is not an isolated and narrow phenomenon. This is rather the essence of genetic processes and life itself. A steady but persistent shift from clear and logically well grounded determinism to indeterminism, which poses serious philosophical questions, does not indicate that science moved from one extreme to another. Science rather adjusted its own position commensurate to a multitude of facts, theoretical interpretations, and philosophical reevaluations. Determinism is not thrown from the pedestal; it has a great deal of merit and can be used widely. Nevertheless, the days of dominant determinism are over. Randomness and uncertainty are no longer poorly grounded scientific views supported by doubtful data. On the contrary these are the frontiers of modern science.

During the last decades Danish theoretical physicist Holger Bech Nielsen developed a concept of "random dynamics" (http://www.nbi.dk/~kleppe/random/qa/qa.html). The objective of this concept is to "derive" all the known physical laws as an almost unavoidable consequence of a random fundamental "world machinery." This sounds quite unusual and even controversial. Smolin (2007) commented regarding these ideas "that the most useful assumption we can make about the fundamental laws is that they are random."

Summary

Life is beautiful and harmonious and most people share this perception. What is the cause of such beauty and order? Charles Darwin answered this hard and probably one of the most central scientific and philosophical questions. He came to the conclusion that there are two forces of paramount importance. One of them is random hereditary variation, which is a wild and blind force of nature. Another force is natural selection, capable of transforming this randomness into adaptations and harmony. Now 150 years after this incredible revelation and countless attempts to test and even to disprove the Darwinian view, this is the most convincing and factually supported theory of evolution. Certainly the time was not spent in vain. The progress in understanding of intricate life processes was amazing and this is very much relevant to both random hereditary variation and natural selection.

Regardless whether Darwin intended to start a scientific and philosophical revolution, he did it. As we all know, stochastic ideology in a clearly stated form was brought to science by physics and this is an exceptional achievement. A minority would think that this ideology can be easily traced back to the middle of the nineteenth century when two giants, Darwin and Mendel, demonstrated the critical importance of randomness. Acceptance of randomness was an act of scientific bravery because it drew the borders of knowledge and indicated the limits of the unknowable. The classical deterministic approach in science, which presumes continuous and unrestricted expansion of knowledge in all directions, was profoundly questioned.

Albert Einstein, being one of the most influential scientists and deep thinkers in human history, nevertheless was not eager to depart from this classical view of science. The overwhelming belief in "universal causation" cost him dearly and he was left among a small minority of physicists who could not accept quantum theory. Scientific dramas of a similar kind but on a smaller scale occur quite often but are rarely reported. The problem is in two classical scientific principles; every event has a cause and the same cause leads to the same effect. Although these principles work well in many situations, they are not universal due to the existence of randomness, which is an undeniable and basic fact of life. The contradiction between the universal causation and ubiquitous randomness probably reflects the current state of affairs. Hopefully some logical and methodological rectifications will be sufficient to overcome this alleged problem.

Following Monod we can also say that chance is the source of all true novelties in living matter. In other words chaos is responsible for novelties. However two other forces that are always at work, self-organization and natural selection, filter and transform innumerable chaotic changes into order. These are two quintessential properties of life: the constant flow of chaotic changes and their selection and coordination. Thus, making uncertainty certain and even beautiful is what life is all about.

References

Barabási, A.L. 2003. *Linked*. New York: Plume.

Buhl, J., D.J. Sumpter, I.D. Couzin, J.J. Hale, E. Despland, E.R. Miller, and S.J. Simpson. 2006. From disorder to order in marching locusts. *Science* 312:1320–1322.

Camazine, S., J.-L. Deneubourg, N.R. Franks, J. Sneyd, G. Theraulaz, and E. Bonabeau. 2003. *Self-organization in biological systems*. Princeton, NJ: Princeton University Press.

Camus, A. 2004. *The fall*. New York: Alfred A. Knopf/Borzoi Book.

Crick, F. 1988. *What mad pursuit. A personal view of scientific discovery*. New York: Basic Books.

Dawkins, R. *The Selfish Gene*. Oxford, UK: Oxford University Press.

de Duve, C. 2007. Chemistry and selection. *Chemistry & Biodiversity* 4:574–582.

Dennett, D.C. 1995. *Darwin's dangerous idea*. New York: Simon and Schuster.

Eigen, M., and P. Schuster. 1977. The hypercycle: A principle of natural self-organization. Part A: Emergence of the hypercycle. *Naturwissenschaften* 64:541–565.

Einstein, A. 2006. *The world as I see it*. Filiquarian Publishing.

Fisher, R.A. 1930. *The genetical theory of natural selection*. Oxford, UK: Clarendon Press.

Gould, S.J. 1989. *Wonderful life*. New York: Penguin.

Hawking, S. *Does God play dice?* Public lectures. http://www.hawking.org.uk/lectures/dice.html (accessed September 17, 2008).

Hegel, G.W.F. 1874. The Logic. In *Encyclopaedia of the Philosophical Sciences,* 2nd edition, §§107–111. London: Oxford University Press.

Interview with Daniel Dennett http://www.pbs.org/wgbh/evolution/library/08/1/text_pop/l_081_05.html (accessed August 16, 2008).

Ishii, Y., M. Nishiyama, and T. Yanagida. 2004. Mechano-chemical coupling of molecular motors revealed by single molecule measurements. *Current Protein & Peptide Science* 5:81–87.

Ishii, Y., Y. Taniguchi, M. Iwaki, and T. Yanagida. 2008. Thermal fluctuations biased for directional motion in molecular motors. *Biosystems* 93:34–38.

Kauffman, S. 1993. *The origins of order*. London: Oxford University Press.

Kauffman, S.A. 1995. *At home in the universe: The search for laws of complexity*. New York: Viking.

Kinghorn, B.P. 1985. Modelled relationships between animal size and production efficiency. *Zeitschrift für Tierzüchtung und Züchtungsbiologie* 102:241–255.

Kurakin, A. 2005. Self-organization vs Watchmaker: Stochastic gene expression and cell differentiation. *Development Genes and Evolution* 215:46–52.

Lee, D.H., K. Severin, Y. Yokobayashi, and R. Ghadiri. 1997. Emergence of symbiosis in peptide self-replication through a hypercyclic network. *Nature* 390:591–594.

Mayr, E. 2004. *What makes biology unique?* Cambridge, UK: Cambridge University Press.

Monod, J. 1971. *Chance and necessity: An essay on the natural philosophy of modern biology* (*Le hasard et la nécessité*). New York: Alfred A. Knopf.

Morowitz, H., and E. Smith. 2007. Energy flow and the organization of life. *Complexity* 13:51–59.

Nowak, M.A., and H. Ohtsuki. 2008. Prevolutionary dynamics and the origin of evolution. *Proceedings of the National Academy of Sciences USA* 105:14924–14927.

Ohta, T. 2002. Near-neutrality in evolution of genes and gene regulation. *Proceedings of the National Academy of Sciences USA* 99:16134–16137.

Orgel, L.E. 2000. Self-organizing biochemical cycles. *Proceedings of the National Academy of Sciences USA* 97:12503–12507.

Parmon, V.N. 2008. The prebiotic phase of the origin of life as seen by a physical chemist. In *Biosphere origin and evolution,* ed. N. Dobretsov, N. Kolchanov, A. Rozanov, and G. Zavarzin, 89–102. New York: Springer Science+Business Media.

Peitgen, H.-O., H. Jürgens, and D. Saupe. 2004. Chaos and fractals. New frontiers of science, 2nd edition. New York: Springer Science +Business Media.

Raser, J.M., and E.K. O'Shea. 2004. Control of stochasticity in eukaryotic gene expression. *Science* 304:1811–1814.

Rashevsky, N. 1966. Physics, biology, and sociology: A reappraisal. *Bulletin of Mathematical Biology* 28:283–308.

Richardson, R.C. 2001. Complexity, self-organization and selection. *Biology and Philosophy* 16:653–682.

Sella, G., and A.E. Hirsh. 2005. The application of statistical physics to evolutionary biology. *Proceedings of the National Academy of Sciences USA* 102:9541–9546.

Shmulevich, I., S.A. Kauffman, and M. Aldana. 2005. Eukaryotic cells are dynamically ordered or critical but not chaotic. *Proceedings of the National Academy of Sciences USA* 102:13439–13444.

Smolin, L. 1997. *The life of the cosmos*. New York: Oxford University Press.

Smolin, L. 2007. *The trouble with physics*. Boston: First Mariner Books.

Stamos, D.N. 2001. Quantum indeterminism and evolutionary biology. *Philosophy of Science* 68:164–184.

Theise, N.D. 2006. Implications of "postmodern biology" for pathology: The cell doctrine. *Laboratory Investigation* 86:335–344.

Theise, N.D., and R. Harris. 2006. Postmodern biology: (Adult) (stem) cells are plastic, stochastic, complex, and uncertain. *Handbook of Experimental Pharmacology* 174:389–408.

Wikipedia. Crystallization. http://en.wikipedia.org/wiki/Crystallisation (accessed September 18, 2008).

Wikipedia. Evolutionary algorithm. http://en.wikipedia.org/wiki/Evolutionary_algorithm (accessed September 16, 2008).

Wikipedia. Logistic map. http://en.wikipedia.org/wiki/Logistic_map (accessed September 18, 2008).

Zuckerkandl, E. 2002. Why so many noncoding nucleotides? The eukaryote genome as an epigenetic machine. *Genetica* 115:105–129.

Zuckerkandl, E. 2005. "Natural restoration" can generate biological complexity. *Complexity* 11:14–27.

Index

Note: Page numbers with "n" after them indicate a note. Page numbers in **bold** refer to figures or tables.

N

For Product Safety Concerns and Information please contact our EU representative GPSR@taylorandfrancis.com Taylor & Francis Verlag GmbH, Kaufingerstraße 24, 80331 München, Germany